Engineering Principles in Everyday Life for Non-Engineers

Synthesis Lectures on Engineering

Each book in the series is written by a well known expert in the field. Most titles cover subjects such as professional development, education, and study skills, as well as basic introductory undergraduate material and other topics appropriate for a broader and less technical audience. In addition, the series includes several titles written on very specific topics not covered elsewhere in the Synthesis Digital Library.

Style and Ethics of Communication in Science and Engineering
Jay D. Humphrey and Jeffrey W. Holmes
2008

Introduction to Engineering: A Starter's Guide with Hands-On Analog Multimedia
Explorations
Lina J. Karam and Naji Mounsef
2008

Introduction to Engineering: A Starter's Guide with Hands-On Digital Multimedia and
Robotics Explorations
Lina J. Karam and Naji Mounsef
2008

CAD/CAM of Sculptured Surfaces on Multi-Axis NC Machine: The DG/K-Based
Approach
Stephen P. Radzevich
2008

Tensor Properties of Solids, Part Two: Transport Properties of Solids
Richard F. Tinder
2007

Tensor Properties of Solids, Part One: Equilibrium Tensor Properties of Solids
Richard F. Tinder
2007

Essentials of Applied Mathematics for Scientists and Engineers
Robert G. Watts
2007

Project Management for Engineering Design
Charles Lessard and Joseph Lessard
2007

Relativistic Flight Mechanics and Space Travel
Richard F. Tinder
2006

Engineering Principles in Everyday Life for Non-Engineers
Saeed Benjamin Niku

ISBN: 978-3-031-79371-4 paperback
ISBN: 978-3-031-79372-1 ebook

DOI 10.1007/978-3-031-79372-1

A Publication in the Springer series
SYNTHESIS LECTURES ON ENGINEERING

Lecture #26
Series ISSN
Print 1939-5221   Electronic 1939-523X

Engineering Principles in Everyday Life for Non-Engineers

Saeed Benjamin Niku

California Polytechnic State University
San Luis Obispo

SYNTHESIS LECTURES ON ENGINEERING #26

ABSTRACT

This book is about the role of some engineering principles in our everyday lives. Engineers study these principles and use them in the design and analysis of the products and systems with which they work. The same principles play basic and influential roles in our everyday lives as well. Whether the concept of entropy, the moments of inertia, the natural frequency, the Coriolis acceleration, or the electromotive force, the roles and effects of these phenomena are the same in a system designed by an engineer or created by nature. This shows that learning about these engineering concepts helps us to understand why certain things happen or behave the way they do, and that these concepts are not strange phenomena invented by individuals only for their own use, rather, they are part of our everyday physical and natural world, but are used to our benefit by the engineers and scientists. Learning about these principles might also help attract more and more qualified and interested high school and college students to the engineering fields. Each chapter of this book explains one of these principles through examples, discussions, and at times, simple equations.

KEYWORDS

engineering concepts, entropy, thermodynamics, thermodynamic cycles, combined cycle power generation, moments of inertia, stepper motors, DC motors, AC motors, transformers, engines, rotary engines, 2-cycle engines, hybrid cars, vibrations, natural frequency, hearing, guitars, signal transmission, Coriolis acceleration, vectors, weather systems, electromagnetic force, EMF, back-EMF

Dedicated to Shohreh, Adam, and Alan
for their patience with me

Contents

Prologue

Almost every aspect of our lives is governed or affected by some engineering concept. Most people do not know about these concepts or are oblivious to them. We take it for granted that at any time, we have cold drinks and safe foods in our refrigerators. This was not true a few decades ago. The principles on which refrigeration is based have existed forever, but we did not know how to use them properly. We take it for granted that wherever we go, a mixture of oxygen and nitrogen is present. We do not suffocate from a lack of oxygen in one location while burning in pure oxygen in another. And we take it for granted that we age—but why? We can walk for hours without really getting tired, but cannot run for long without getting tired. Why is it that cities on the Atlantic coast get snow, but the ones on the Pacific coast do not? Tree branches get smaller in diameter as their distance to the trunk increases, but why? And a telephone book can be easily bent, but not a piece of cardboard. Why? Most hybrid cars have better gas mileage in the city than on freeway driving. Why do airplanes fly, how do engines work, and how do we hear? Why is it that we can select the broadcast from one station at a time without mixing information from hundreds of others?

So how are these issues related to engineering (or as some would call it, physics)? Simple principles govern why and how things happen. If we learn these principles we can understand how different phenomena affect our daily lives, why certain things happen as they do, and how we can use them to our benefit. There are too many engineering principles to discuss for non-engineers. This book covers a few of these principles that are more general and directly apply to our understanding of natural phenomena. A few equations are used in the discussion to better understand the issues. I hope this will not be a distraction. I envision that whether or not you are an engineer or scientist, you will be able to follow them.

This book can be a general reference book for learning about engineering for all audiences, especially for college students in majors other than engineering, or used in general education classes for technical content, or for encouraging high school students into thinking about STEM (science, technology, engineering, and mathematics), or general non-fiction reading. Many individuals from all walks of life have made encouraging comments about how they enjoyed reading the manuscript and how they have learned from it, and I thank them for their time.

Not knowing about a subject does not make a person dumb. Many books are supposedly written for dummies. The assumption in this book is that people are intelligent, even if they do not know about certain fields; as long as they have the perseverance and are patient they can learn about new subjects. So this book is not for dummies. It is for intelligent individuals who are interested in learning new material.

I would like to foremost thank Alan Niku for his thorough and thoughtful editing of the manuscript, his comments, his humor, and his photography. I would also like to thank Joel Claypool for his courage in taking on this project and Dr. C.L. Tondo for his work on the project, and Hila Ratzabi for editing the manuscript. In addition, my sincere thanks go to Daniel Raviv, James LoCascio, Patrick Lemieux, Frank Pekar, William Murray, Steven Klisch, Julia Wu, Jesse Maddren, Glen Thorncroft, Ahmad Nafisi, Larry Coolidge, Sina Niku, and others whom I may have forgotten by now, for their assistance during the writing of this project.

Let's get to work. We will look at a few concepts and see how they relate to our lives every day. I hope you will enjoy the book and learn something new from it.

Saeed Benjamin Niku
Mechanical Engineering
Cal Poly, San Luis Obispo
January 2016

CHAPTER 1

Entropy

Natural Orders, Thermodynamics, Friction,

Hybrid Cars, and Energy

1.1 INTRODUCTION

Imagine walking into a room and finding out that all of the oxygen in the room has separated into one side and all the nitrogen and other gases in the opposite side, and consequently you can either not breathe or you almost get burned in the pure oxygen. Certainly this does not happen in nature. In reality, even if you design a chamber with a membrane in the middle, fill one side with oxygen and the other side with nitrogen, and then pull out the membrane, the oxygen and nitrogen completely mix after a while. The natural world does not like this artificial order; thanks to *entropy*, it will mix them together into a uniform mixture. Unless other forces and characteristics (such as differences in density, non-mixing fluids) intervene, things get mixed up into uniform states.

In fact, this is true for any artificial order created against the natural disorder of things. We may build a solid structure; nature will destroy it one way or another. We come into living; nature will eventually find ways to kill us. Mountains and valleys are formed by other forces of nature; the mountains eventually wash off into the valleys to fill them into a natural disorder or uniformity. Boiling water will cool to the same temperature as the environment. Only natural orders remain, to some extent.

So what is *entropy*?

Entropy is a measure of the level of organization (or disorganization) in a system; a degree of its organization or randomness; the probability of knowing where the molecules of a medium are at any given time. When order is increased and the system becomes more organized, entropy decreases (yes, decreases). When order is decreased, entropy increases. In all natural systems, there is a tendency for entropy to increase. When an artificial order is imposed on a system the entropy decreases. Can entropy eventually increase to infinity? Perhaps. But we know that it increases when order is reduced.

In the following examples, we will investigate the role entropy plays in reducing order.

Example 1.1 Ice in Warm Water

Imagine a glass of water at your bedside overnight. If you measure the temperature of the water in the morning, it will be the same as the room's temperature; there is no difference between these temperatures, and therefore, no heat is transferred from one to the other. Even if the water was originally warmer than the air, or colder than it, after some time, heat would have transferred from one to the other to bring both media to the same temperature.

Now drop an ice cube into a glass of warm water. Similarly, since there is a temperature difference between the ice and the water; heat is transferred from the water to the ice until they are both at the same temperature. The concentrated amount of thermal energy in the water compared to the ice is not natural; in natural systems, heat is transferred in the direction of higher-to-lower temperature from one system to another until they are all at the same temperature, whereas here, the ice is cold and the water is warm. Therefore, there is an "un-natural" order in this system. We have deliberately created a system that has a particular order to it; entropy will destroy this order through the transfer of heat between them. Not only is it impossible to wake up one day and see that the glass of water you left at your bedside has transformed itself into a glass of warmer water plus an ice cube in the glass (by transferring heat from a portion of the water into the rest to make an ice cube), even if you create this order, entropy will nullify it.

The same is true in other systems. Cold air versus warm air, a hot plate and a cold dish on it, a stove and a pot of water, a heater and the cold air surrounding it, and the air-conditioned building and the warm weather outside. In all these examples, heat flows in the direction of higher-to-lower temperature until they are the same as the temperature of the place in which they are and achieving equilibrium, therefore eliminating the order caused by the difference in temperatures.

Example 1.2 Mountains and Valleys

Mountains are above the average horizon and valleys are below, therefore creating an order. Even though mountains and valley are themselves created by natural forces, this difference creates a particular un-natural order. Through different mechanisms, nature will try to break down the mountains and fill up the valleys to destroy this order, even if it takes millions of years. We cannot expect the dirt to simply rise to create mountains in the opposite direction.

What are some of these mechanisms?

As air is heated and the energy of its molecules increases, the distance between these molecules increases too, making the air less dense. As a result, the warmer (and less dense) air rises. Since entropy does not favor the creation of a new order (leaving empty the space that the warm air previously occupied), cold air moves in to fill the empty space, thus creating winds and currents. The wind moving through the mountains erodes the mountains ever so slowly and moves the material away to reduce it. To this, we should also add flowing waters when it rains and snow melts, or when rivers cut through the dirt and carry it along. The Nile River is famous for flooding at its delta every year and fertilizing the area with new and nutrition-filled dirt from the mountains above. Figure 1.1 shows the erosion of earth due to these mechanisms at work at Petra.

Figure 1.1: The washed out, worn out Petra valley.

Most materials expand when heated and contract when cooled, including when frozen. This is due to the fact that as the temperature decreases, so does the energy of the molecules, and therefore, they get closer to each other, reducing the volume of the material and making it denser. However, exceptions exist, including Bismuth and water; they expand when frozen. The density of water is maximum when it is at about 4°C at sea level. The volume of water increases when it freezes and when it is heated; this is why you should not place a closed water container in the freezer; it may rupture. You can see the result of this expansion by observing the surface of an ice cube. If you mark the surface of water in a plastic container and place the container in a freezer, you will notice that when frozen, the surface of the ice will have risen above the original line. However, if you use a metal container (and cover the surface with a piece of cardboard to prevent the direct blowing of cold air onto the surface) you will notice that the surface of the ice is somewhat conical in shape (although shallow), as in Figure 1.2. In most refrigerators, the cold air is blown throughout the freezer compartment. When a plastic container is used, since it is not a very good heat conductor, the water freezes at the surface first, and therefore, as the water expands, it pushes up the surface as it continues to freeze. However, since the metal container is a good conductor, it causes the water to start freezing at the perimeter of the surface and around the walls. As it continues to freeze and expand, the surface rises slowly and continues to freeze at slightly higher levels until it finishes with an apex in the middle, just like a hill (the complete mechanism at work here is more complicated though).

Water pipes may also rupture in the winter when they freeze. In places where the winter temperatures are very low, water pipes are buried a few feet underground to keep them warmer

Figure 1.2: In a metal container, as water freezes around the perimeter towards the center, it expands and pushes the water level up slowly, creating a conical shape with an apex.

and prevent their rupture. The same is true for the water that gets in between crevices of rocks and other rigid material, whether on a mountain or not. When the water freezes, it cracks the rock or breaks it apart, and therefore, eventually destroys the order that exists.

And what happens to the tree branch that falls in the forest when a hurricane passes through or when the tree is diseased? Microorganisms, beavers, termites, and other agents eventually break it down and destroy it.

Example 1.3   What Happened to the Flagpole?

So, you had a flagpole in the yard, and after a few years, it rusted and eventually failed. The same is the fate of the sign pole at the bus-stop. They rust because there is an order in the system that is un-natural. Entropy demands that the order be destroyed. One may try to postpone the rusting by painting the pole or embedding it in concrete, or keeping away the moisture whenever possible. But rusting, or oxidation, is another natural mechanism for reducing the artificial order of things. Just think about the radiator in your car. It will eventually rust and fail.

But then what shall we say about the steel used in an engine, where it is constantly lubricated and moisture is kept away? It hardly rusts. In this case, friction and rubbing action between the different moving parts will eventually erode the material, eventually reducing it into nothing (although the engine will stop long before this state is reached, and therefore, oxidation takes over in the junkyard).

So do all metals oxidize? Not really. Stainless steel is almost rust-proof (although not all types and all qualities are rust-proof. Less expensive stainless steel that lacks enough chromium will rust. Check out your barbeque). Gold does not rust either, but it wears out. So it is not impervious to entropy. And what happens to that plastic part that does not rust? The ultraviolet light will help in its decomposition, first by fading, then hardening and cracking, and eventually decomposing. Is this not what happens to most paints too?

Nature, in its arsenal of mechanisms for reducing order, has countless other weapons, including insects, animals, microorganisms, viruses, diseases, fire, wind, floods, and many others, each with a unique capability to do its work.

1.2 ENTROPY AND EFFICIENCY

Reducing entropy will increase the efficiency of a system by increasing the order within the system. To understand this, let's look at the following case.

Example 1.4 Going to the Office

Naturally, one may need to sleep longer one day due to being tired, less another day due to sickness or having to take care of a chore. It seems more natural that people who all participate in similar activities in the same place like an office or classroom or a factory may like to go to work or to class when they feel like it or when they can. Wouldn't it be nice to be able to go to a classroom any time you prefer? Except that such a system is not efficient. A teacher would have to repeat the same material countless times every day as students showed up randomly when they wanted. One might go to a bank for a transaction. However, if the tellers and bank managers came to work as they wished, chances that the work got done would be very low. Meetings would never work either if participants would arrive when they wished. Think what would happen if the workers of a factory took vacations as they wished instead of all taking the same days off. The factory would not be efficient either.

However, we will create an un-natural situation by requiring that everyone, regardless of how long they have slept, whether they are tired or not or whether they like it or not, must go to a class or be present in the office or participate in a meeting at the same designated time. But as a result, the teacher will have to teach once, the customer can make a transaction with the office staff, and everyone knows when they can expect to see someone for an appointment. Is this not more efficient? Yes, but it is not natural. The tendency in this system is also to reduce order; people skip work, get sick, do not come to work on time, and appointments are not kept. To maintain efficiency, we need to sacrifice comfort or desires. Otherwise, natural chaos would abound.

1.3 IS IT POSSIBLE TO DEFY ENTROPY?

Yes. Simply by creating order, we defy entropy. By doing so we increase order, and consequently, efficiency. As a result, we create systems that do things for us, from which we benefit even if we need to constantly fight entropy.

As an example, take the engine of a car. The design of an engine forces a particular sequence of events to repeat thousands of times a minute for hundreds of thousands of miles of travel. Each time the air is sucked into the cylinders, it is compressed, fuel is injected into the hot air causing an explosion and creating useful work that turns the engine, and the burnt fuel-air gases are forced out (see Chapter 4). This is not a natural sequence and does not happen by itself, but precisely because of that, it is efficient (useful to us). Entropy tends to create situations to reduce this order when components break, rust, or wear out and as accidents happen. But as long as the order is maintained, the system is a useful entity.

If you think about any other system, including living systems, the fundamentals remain the same. A completely chaotic system is natural. However, everything has a particular order in it that makes it useful. We, and everything we create, are under the control of this fundamental phenomenon, even if we learn to defy it.

1.4   WHY DO WE GET OLDER?

As mentioned previously, living systems are also subject to the same entropy. Humans, for example, are very sophisticated and orderly systems. Nature, based on entropy, uses a variety of mechanisms through which it destroys this order, including disease, accidents, and aging. We get older because our systems have to eventually stop. Obviously, here I will not discuss the inherent mechanisms that the body uses to cause aging (such as the natural DNA markers that measure our age and turn on or off different functions of our DNA). Whatever these mechanisms, they are just tools through which entropy does its work. However, even if we do our best to take care of our bodies and stay healthy, even if we never contract a disease, never smoke or drink, always exercise, eat right, and always make prudent decisions, we still get old and our bodily functions eventually stop, some sooner, some later.

Physiologists and engineers can derive equations that describe this phenomenon mathematically as well. For example, considering caloric intake (how much energy is consumed by an individual through the food the person eats) versus the expenditure of energy by a person, one can calculate the efficiency of the human body and the food eaten. Of course, the efficiencies of different food systems are different. However, this waste of energy can be mathematically related to entropy generation. Therefore, one can actually estimate the increase of entropy of the system of human body.

Now think about the fact that when a child scratches her knee, when a person cuts his finger while cooking, or even after surgery, the body heals itself. But why should it? Is that not against entropy? Yes it is. When it heals, the body adds to its order, reducing entropy. So healing, and in fact life itself, are against entropy. As long as we are alive, our bodies have learned to defy entropy and to heal, overcome diseases, grow bigger, re-produce, and create new life. Each one of these phenomena is against the fundamental principles of entropy, disorganization, and the randomness of nature. But we are born (regardless of our will and against entropy), we grow bigger and taller and stronger, we get better after illness, and we go on living for a period of time. Each one of these reduces the overall entropy of the universe until we eventually die, when it increases once again. Planting a seed and growing it into a tree is the same. The secret of life in the seed can defy entropy, causing it to sprout, grow, create fruits and seeds, and resist death for a while. But eventually, all vegetation and trees come to an end too.

1.5 ENTROPY AS DESCRIBED BY AN EQUATION: THERMODYNAMICS

Thermodynamics is the study of thermal systems, resulting in the transformation of different forms of energy, the creation of useful work, heat transfer, and the dynamics involved. Thermodynamics is built on two basic laws, appropriately referred to as the *first law* and the *second law*. (There is also a *zeroth law* that indicates that if two bodies are at thermal equilibrium with a third body, they are also at thermal equilibrium with each other. Kind of obvious, but necessary.) There is also a third law of thermodynamics, stating that entropy is zero at absolute-zero temperature.

In any system, both the first and second laws of thermodynamics must be satisfied, not just one.

1.5.1 THE FIRST LAW

The first law of thermodynamics relates to the fact that energy is not created, but only transformed from one form to another. Therefore, for every system (we refer to it as a closed system, separated from the environment or other systems), the total energy remains the same because we cannot produce any energy and we cannot destroy any energy, but only transform it into other forms. For example, in our cars, we transform into mechanical energy some of the chemical energy stored in the fuel by first transforming it into thermal energy (combustion). The remaining part of the energy is rejected from the engine through the exhaust and the radiator (if water-cooled) and by direct radiation of heat into the atmosphere. The mechanical energy of the engine is also transformed into other forms, for example kinetic energy of the car moving at a particular speed, or the potential energy of the car going uphill (we will talk about these later). This potential energy is once again converted to more kinetic energy (going faster) if we continue downhill. All of this energy is eventually converted into thermal energy through the brakes, air resistance, and friction in the system (we eventually slow down and stop).

The total amount of chemical energy going to the engine is equal to the total mechanical energy plus the rejected heat. This can be described by:

$$E_{fuel} = E_{mechanical} + E_{rejected} \tag{1.1}$$

or

$$E_{in} = E_{used} + E_{rejected}, \tag{1.2}$$

where E_{in} is the input energy, is the E_{used} energy that is converted into a useful form such as mechanical energy, and is the $E_{rejected}$ thermal energy rejected to the environment. You might see this equation containing the word *work* as well, such as *the work done by the engine*. Work is another form of expressing energy. When a force moves, it works. Therefore, when the force created by an engine at the point of contact between the wheels and the ground pushes the car forward, it is doing work. Work, which can be transferred into mechanical energy, can be calculated by multiplying the force by the distance travelled or a torque by the angle rotated. This is usually expressed as:

$$W = F \cdot d, \tag{1.3}$$

where F is the force and d is displacement (how much the object has moved).

This is true for the human body too. If you consider the human body as a system, then the total energy intake (from the foods we eat) should be equal to the generated work, the stored energy in the body (such as by fat), and the rejected heat. So imagine that a person takes in 2,000 Calories of energy in one day. Suppose that the person uses 900 Calories to walk, think, perform bodily functions, talk, etc. Also imagine that the person loses 900 Calories of heat through radiation and convection. Our bodies, when warmer than the environment, lose thermal energy. We must lose the thermal energy generated by our muscles and physiological functions to not only remain comfortable, but to even stay alive. The opposite is true in temperatures warmer than our body temperature; not only does the body get warmer, it cannot reject its extra thermal energy through convection or radiation. The reason we perspire is to increase heat loss from the body by evaporating the sweat on our skin (it takes thermal energy from the body to evaporate, therefore transferring our body heat into the environment. In damp environments where humidity is high and the body cannot easily evaporate the sweat, we feel warmer. Similarly, when there is a breeze, we feel cooler because it increases heat loss from the body). Without this heat loss, we may die (should we use anti-perspirants in hot weather?). The remaining 200 Calories that are not rejected and are not used otherwise will be stored by our bodies in the form of fat. At 9 Calories per gram of fat, this is about 22 grams of fat added to the body, a weight gain. On the other hand, if the energy intake is less than the work produced plus the heat loss, our bodies will convert the body fat into energy supply, therefore causing a net weight loss. The energy equilibrium requirement is maintained through this dynamic.

It should be mentioned here that this is a very simplified model of the human body. In general, each person has a different metabolic rate that is affected by many factors, including heredity, age, and so on. Some of the food we eat is not digested at all, and is rejected as waste. When we suddenly reduce our energy intake (say by dieting), the body assumes there may be a supply problem, like a famine, and slows the metabolic rate, storing more fat for future emergencies. If we start eating more again, it will convert more to fat. If we eat as usual but work more (burn more calories), there is less left for conversion to fat. Therefore, the aforementioned model should be used for understanding the energy equilibrium and not a complete picture of human metabolism or dieting. In general, reducing energy intake should have the same effect as doing more work (more walking, exercising, swimming, etc.). However, doing more activities, even in light of eating more calories, is more fun!

By the way, you may have noticed that in the preceding section, we used Calories and not calories. Each Calorie is in fact one kilocalorie or 1,000 calories. In food science notation, the unit used is Calorie. It is generally assumed that carbohydrates and proteins are approximately equivalent of 4 Calories per gram, or 4,000 calories per gram. Fat is approximately 9 Calories per gram, or 9,000 calories per gram. One calorie is the energy needed to raise the temperature of 1 cc (cubic centimeter) of water by 1 degree Celsius (or 1.8 degrees Fahrenheit). Therefore, one Calorie is the energy needed to raise the temperature of one liter of water by one °C. Consequently, 2,000 Calories consumed by one person in one day can heat up 2,000 liters of water by one °C. A person requiring 2,000 Calories per day for maintaining constant weight, who fasts for 24 hours (no food intake whatsoever, except water), and who still performs routine work as everyday life that burns 2,000 Calories of fat, will lose 2,000/9 = 222 grams or a little less than 0.5 pounds. Now calculate how many days one would have to fast, completely, and maintain the same level of activity, to lose a desired amount of body fat.

So, can we cool down a room during the heat of the summer by placing an air-conditioning unit or a refrigerator with its door open inside the room? Do either of them not produce cooler air that can make the room cooler? The answer is no. The room will actually be hotter. This is because there is friction in every system, regardless of what we do. So we need energy to overcome this friction. We also need energy to transfer the heat from one part of the system to another. Remember, energy is neither produced, nor destroyed, but only transferred (or converted) from one form to another. The cooler air of the refrigerator or the air-conditioning unit is the result of a thermodynamic cycle (we will discuss this in Chapter 4) which removes the thermal energy through one part of the system called an *evaporator* (and therefore, making that part cooler) and transferring it to another part of the system through the *condenser* (and therefore, making it hotter). The net result is that we have spent energy in order to do this transfer. If the evaporator part of the system is outside of the room, and therefore, transfers the additional thermal energy to the ambient air, the net result is that the room or the refrigerator will be cooler inside at the expense of being hotter outside. If the whole system were within the room, the net result would be a hotter room. Note how we have created a particular order within this system that against the expected outcome (that heat flows from a hotter place to a colder place) transfers the heat from a colder place to a hotter place by the additional work we do.

It is necessary to mention one thing here. If you look at a dictionary, the word *heat* is defined in terms such as "energy associated with the motion of atoms or molecules in solids and capable of being transmitted by convection and radiation," as a "form of energy possessed by bodies," the "perceptible sensible or measurable effect of such energy so transmitted," etc. In vernacular conversations we also refer to heat as energy. However, in thermodynamics, heat is the transfer of energy from one medium to another, which is different. Although in the realm of thermodynamics equating these to each other is not correct, we still use the word as if it is an energy term and refer to heaters, heat exchangers, heat pumps, heat absorption and heat rejection, and similar terms. Inadvertently, the word is also used here as if it were energy because we normally refer to it as such. Understanding that although correct definitions are different, we may sometimes use the word "heat" as if it were thermal or internal energy.

1.5.2   THE SECOND LAW

The second law of thermodynamics relates to the quality of this energy transformation. But first a word about energy types.

Rub your hands against each other for a few seconds. They will feel warm. Burn a small stick of wood. It will give off thermal energy. Turn on a light bulb (especially an incandescent light bulb that gives off light through a heated element) and it gets hot. Run the engine of your car, and it too will get hot. Now try to do the opposite: use the heat to move your hands, to recreate the stick of wood, regenerate the electricity that turned on the light, or recreate the gasoline that went into the engine. You will need to create a system composed of many elements to move your hands, spend the energy for a long time to nurture and raise a tree, make a power-plant or design a device like an engine, or use a chemical reactor to re-make the gasoline. This is because thermal energy is the lowest-quality energy. All other forms of energy tend to reduce to thermal energy unless we do something drastic. Natural systems go toward the lower-quality thermal energy. For example, what happens to the energy of your voice as you speak? Your voice will vibrate countless different systems in your vicinity, including surfaces and molecules of air through which the energy eventually converts to thermal energy. In fact, the sound can only emanate in air by vibrating the molecules of air; sound does not transfer in vacuum. And all that mechanical energy in the form of vibrating elements converts to thermal energy. And what happens to the sound level if you speak to someone while you are inside and the other person is outside of a room? The reason the level of your voice heard outside is lower is that part of the energy is absorbed and converted into thermal energy by the walls, the doors and the windows, and the furniture and other things in the room.

Additionally, the efficiency of converting other energy forms into thermal energy versus converting thermal energy into other forms of energy is very different. Not only is it easier to convert electrical energy to thermal energy, it is also more efficient. Therefore, the best efficiency one might expect from a power plant that converts lower-quality thermal energy (chemical energy of the fuel converted to thermal energy during burning) into higher-quality electrical energy is about 40% (see the discussion about "*combined cycles*" in Section 4.8), whereas converting electrical energy into thermal energy is very efficient (almost all electrical energy is converted into thermal energy in a lamp). This is also very much related to entropy, but we will not get into it for now. The efficiency of converting electrical energy into mechanical energy (such as in an electric motor) can be over 90%. Similarly, electrical-to-electrical conversion of energy such as in a transformer or a charger can also be about 90% or so.

Another important issue is the value or utility (usefulness) of energy. What matters is not only the total amount of energy that is available, but also at what temperature it is (this is called *exergy*). A high temperature medium is higher in value or utility than the same medium in low temperature. For example, the total energy stored in the waters of a lake is tremendous, but since its temperature is about the same as the ambient temperature, this energy cannot readily be used. A small mass of fluid at high temperature may have the same energy as a large tank of the same fluid at ambient temperature. We cannot easily use the energy of the larger mass at near ambient temperature, but the energy of the high-temperature small mass can be used readily (for example, to heat a glass of water) and, therefore, it has more utility.

As mentioned earlier, both the first and second laws of thermodynamics must be satisfied. Imagine a cup of hot coffee left in a room. Eventually, the heat transfers from the coffee into the room, and therefore, the energy lost from the coffee is gained by the room, satisfying the first law. However, if it were up to the first law alone, it should also be possible that some energy from the room transfers itself to the now-cooled cup of coffee and heats it again; if it were left to the balance of energy transfer alone, both scenarios would satisfy the first law and either one would be possible. But we know this does not happen, because it violates the second law. Based on the second law of thermodynamics, it is impossible for the heat to transfer from a cold source into a hot source on its own.

A very common blessing and curse of everyday life is friction. Friction is a blessing when we need it, and a curse when we do not need it. Examples abound, but for instance, when we brake, it is friction that stops our car or bicycle. In this case, more friction generally makes a better brake. The same is true in walking; we can only walk because there is friction. Just think of walking on ice or with roller skates and how hard it is even though the friction is low, not zero. And we can grab a fork and eat only because there is friction. In all these cases we win because there is friction. But we lose when there is unwanted friction, for example in the mechanical components of our car, air friction (drag) as we move through the air, and friction on the floor when we push heavy objects. However, in both cases, whether a blessing or a curse, friction always opposes motion and therefore it always converts part of the energy (or all of a specific form of energy such as kinetic energy) in the system into thermal energy, the lowest form of energy. Since every real system has friction, every real system creates wasted heat, whether a car, a fan, a computer, or our bodies. There is no escape from this. Therefore, as every system operates, it will lose some of its energy into thermal energy, and consequently, there can never be a 100% efficient system (this is also directly related to entropy and is expressed as a thermodynamic equation that is used in the analysis and design of systems, but is beyond the scope of this book).

In fact, based on the preceding argument, perpetual machines are fundamentally impossible. Since every system has friction in it, it is impossible to drive a system indefinitely without supplying some energy into it; the friction in the system converts the added energy into thermal energy. Without it, the machine will not move, and even if the system is given some initial energy (like a flywheel which is already rotating) the stored energy will soon be converted into thermal energy and rejected due to the friction in the system. Based on this fact, next time you hear someone's idea of a novel, innovative, and unique perpetual machine, go ahead and bet that it will not work and challenge them to make it if they insist that you just don't understand.

So what is the second law of thermodynamics anyway? The second law states that the transfer of energy from one system to another is in the direction of lower-quality energy. For example, a hot glass of water left in a room will lose its energy to the cooler air in the room until they are at equilibrium and there is no more transfer of energy. What is the chance (probability) that the energy in the cooler room would somehow collect itself into the glass of water and make it hotter than the room temperature? Zero. Is this not what we already said about entropy anyway?

1.6   HYBRID CARS ANYONE?

Hybrid cars have recently become very popular, and rightly so, due to their very high efficiency as compared to other vehicles with regular internal combustion (IC) engines. For cars of a similar size and weight, typical fuel consumption may be in the 25–30 MPG in the city and 30–40 MPG on freeways, whereas for a hybrid it might be 50–60 MPG in the city and about 40–45 MPG on freeways. Obviously, hybrid cars are much more efficient, even if the numbers are strangely confusing in that hybrids use more gasoline in freeway driving than city driving.

First, let's see why non-hybrid cars are more efficient in freeway driving than in city driving. As previously discussed, the total energy of the fuel spent is converted into the kinetic energy of the car and into thermal energy due to friction, drag, sounds and vibrations, as well as the huge amount of energy rejected by the engine through the exhaust, radiator, and heat loss through the body of the engine (and due to the second law of thermodynamics, it is impossible to eliminate this loss). The efficiency of the best engines is lower than 40%; this means that less than 40% of the total energy is converted into useful energy such as kinetic energy, while the rest is lost as thermal energy. This is even worse when the engine runs without the car moving, such as behind a traffic light or in congested traffic. In these situations, there is no kinetic energy, and therefore all the fuel energy is wasted.

The kinetic energy stored in a car of mass m at a velocity of v is:

$$E = \frac{1}{2}mv^2. \tag{1.4}$$

So, in stop-and-go driving in the city, every time we speed up, we convert less than 30%–40% of the spent fuel energy into kinetic energy. In a non-hybrid car, when we brake, this energy is converted to additional thermal energy and is rejected to the environment, causing more loss. In fact, sometimes the thermal energy of braking can be so much that it may damage the brake's rotor assembly. The rotor gets so hot that, due to what is called *residual stresses* that remain in it as a result of manufacturing operations, it can bend, requiring that it be grinded to prevent pulsations when the brake is applied. The more we accelerate and gain speed and then slow down by braking, the more energy we waste, leading to higher gasoline consumption in the city compared to freeway driving where we drive at a more steady speed. In this case, since we do not slow down or completely stop as much as we do in city driving, there is much less waste, and therefore better efficiency. Engines require an injection of extra fuel to accelerate when we speed up, further reducing the efficiency in city driving. As a result, more uniform speeds in freeway driving increase efficiency, reducing the need for additional gasoline. To make matters even more complicated, since most drivers like to have plenty of power available to accelerate quickly, more powerful engines are installed in cars; in freeway driving, when accelerations are lower (smaller changes in speeds in freeway driving), only a fraction of the available power of the engine is used. Since engines have different efficiencies at different power levels, the total efficiency of the engines is affected by the power it generates.

In purely electric cars (also called *Zero Emission Vehicles* or *ZEV*, *Battery Electric Vehicles* or *BEV*, *Electric Vehicles* or *EV*), instead of an engine, the source of energy is a large set of batteries. The car is propelled by converting the electrical energy stored in the batteries (actually in the form of chemical energy) to mechanical energy through electric motors that rotate the wheels. Like in a conventional car it is possible to stop an electric car by braking, and therefore, converting the kinetic energy of the car into thermal energy and rejecting it into the atmosphere. However, due to a phenomenon called *electromotive force* (or EMF), the kinetic energy can be recaptured and converted back to electrical energy that can be used to recharge the batteries. Of course, this is not 100% efficient either and some energy is lost during conversion (in fact, due to safety concerns, there is always some braking mixed in with regeneration to ensure that the vehicle stops when needed, especially at low speeds). Therefore, since a lot of the kinetic energy is captured and converted back into electrical energy and stored in the batteries (or ultra-capacitors), the efficiency of the system is much higher than when an internal combustion engine is used. As a result, the total efficiency of an electric car can be much better than a regular engine-equipped car.

A Word about Electromotive Force (EMF)

Imagine that a conductor (for example a wire) is placed within a magnetic field (for example between the two poles of a magnet) as in Figure 1.3. With an electric current flowing through it, the conductor will experience a force (or a torque in rotational systems) called electromotive force, which is perpendicular to the plane formed by the current and the field. Similarly, if the conductor is moved (or rotated) within the field (for example, by applying a force or torque to it and causing it to move), a current will be induced in the conductor. The same is true if the magnetic field is turned on and off or is changed in magnitude or direction. This simple principle governs how electric motors work, how electric generators generate electricity, and even how transformers change the ratio of voltage and current for electrical energy distribution systems. More on this in Chapter 6, but for now suffice it to say that an electric motor and an electric generator (such as the generator in your car that recharges your battery after you drain it a bit by starting the engine, and which if it fails, your battery will drain, stranding you in the middle of worst places when you need to restart your engine!) are the same thing (with minor differences for managing the DC and AC currents). Figure 1.4 shows the stator coils of an AC induction motor and its rotor. The rotor is a collection of conductors that move within the magnetic field, generated by the stator coils.

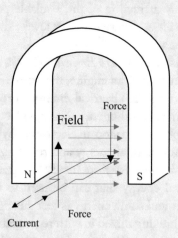

Figure 1.3: A wire carrying a current, placed within a magnetic field, will experience a force in a direction normal to a plane formed by the current and the field.

This means that if a current passes through a motor's coils, the electromotive force will cause it to rotate, acting as a motor. However, if you rotate the shaft of the motor, either by attaching it to something else that is rotating or by turning it manually, it will have a current induced in it, acting as a generator. Figure 1.5 shows a flashlight in which the user can rotate the handle to charge the energy-storage unit within the flashlight. In this case, instead of a rechargeable battery, a large-capacity capacitor may be used.

Figure 1.4: An AC motor with its stator coil and rotor.

Figure 1.5: In this flashlight, a crank is used to rotate a generator in order to convert mechanical energy into electrical energy. This is either stored in a rechargeable battery or in a large capacitor.

However, as explained earlier, the mechanical energy is converted into electrical energy through the generator (which is really a small motor) and stored in the capacitor. Similarly, in a hybrid car, the same motor that rotates the tires when powered by the current from the batteries can also function as a generator and produce a current that recharges the batteries when it is forcefully rotated by the wheels (while the current from the batteries is cut off). Since a torque is needed to turn the generator, it converts the kinetic energy of the car into electricity, and consequently acting as a brake. Obviously, electronic circuits are used to control the flow and the charging of the batteries and how much force is applied.

So, in electric and hybrid cars instead of braking by mechanical means and wasting the energy into thermal energy, the kinetic energy is recaptured and converted back to electric form, stored in the battery, and used again later.

It should be mentioned that, when charged by plugging into the electric grid, the energy stored in the batteries comes from a power plant that also burns fuel, so it is not that much more efficient than an engine; the difference is that power plant energy conversion systems are somewhat more efficient than engines, perhaps a little over 40% or so (see Section 4.8 for additional notes on this). Electricity can be "generated" (or more accurately, converted) by burning gas or coal (less expensive), by solar panels, through nuclear or hydroelectric power plants, wind energy systems, etc. Therefore, the total system is more efficient than an engine.

Electric cars have many advantages in addition to their efficiency. Since there is no engine, there is no need for oil changes, maintaining the water level in the radiator (there is no radiator for cooling the engine anyway), and almost no need for replacing brakes. They also do not have a gear box or a clutch (or transmission). Figure 1.6 shows a Tesla Motor S-Type car without the body. It shows how there are very few parts to the car. However, electric cars have a fundamental drawback. When the battery is drained, it must be recharged, whether at home, at work, outside shopping malls, or at charging stations. Unless the car is driven short distances, for example between your house and your place of work or school or shopping, and time is available to recharge the batteries at night or while you shop, or if charging stations are readily available to charge the batteries on a regular basis (which requires time), you may run out of energy and not be able to drive your car. This severely limits the range of an electric car and limits its usefulness to only short trips. Some companies have proposed, and have attempted at great expense, to create battery-exchange stations as common as gas stations, into which you may drive your car, automatically exchange your battery with a similarly charged unit, pay for the energy used, and quickly get on your way again. Until such a time when there are sufficient stations everywhere, the recharging of batteries will remain an issue.

A huge advantage can be created if a relatively small engine is also added to the purely electric system to recharge the batteries at a constant rate while we drive. These cars are referred to as *Range Extended Electric Vehicles* or *Plug in Hybrid Electric Vehicles (PHEV)*. In this case, the car has a complete electric drive system, including batteries, drive motors and generators for brakes, and control systems, but also an engine, fuel tank, and associated hardware. However, in general, the power required to constantly recharge the batteries is a fraction of the power needed to propel the car at high accelerations. Therefore, a small engine can be designed and used to generate electricity at its maximum efficiency to recharge the batteries. This way, the driver may drive the car just like a regular car without regard to range limitations. Since most of the energy is recaptured during braking, and because they can be charged at night, the efficiency of these cars

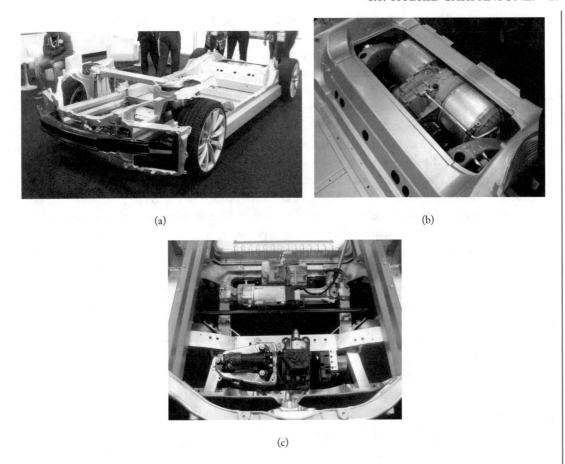

Figure 1.6: Tesla Motors S-Type chassis (a), drivetrain (b), and steering mechanism (c). Due to its nature, an electric car is very simple, with very few parts, compared to a conventional car or a hybrid. The batteries are placed in the middle part of the car under the seats.

in terms of needed gasoline is very large. However, if the intention for the engine is to propel the car in extended driving situations like a regular engine capable of large accelerations, the engine will have to be larger and its efficiency will be lower. For example, the 2012 model Chevrolet Volt has an 84-horsepower engine. The engine of the 2013 Toyota Prius is 138 HP.

So then what is a hybrid electric vehicle (*HEV*)? A hybrid car is the combination (hybrid) of both systems that share the power generation duties. Although it sounds even more inefficient to have both an engine and a set of usually heavy and expensive batteries with limited lives, and drive motors and control systems, hybrids offer something that electric cars lack: the convenience of having gasoline available to burn regardless of how long a drive might be, as well as the desired accelerations available when needed, all at a much better gas mileage. It should be mentioned here that there are many different combinations of gasoline and electric drive duties used in hybrids, each with their own characteristics (for example, parallel systems, series systems, and power-split systems). Regardless, in hybrid cars, when not plugged in to recharge the batteries, a regular engine converts the chemical energy of the gasoline into electrical energy which is used to charge the batteries or assist in powering the vehicle. The electrical energy is used to drive the electric motors to propel the vehicle, and the EMF is used to convert much of the kinetic energy of the car back to electrical energy when we brake. As long as the speed of the car is relatively low (up to about 30–35 miles per hour) and the batteries are charged, the electrical energy is used to drive the electric motors and propel the vehicle. When the battery is drained, the engine automatically starts to charge the battery and also help propel the car. At higher speeds the engine starts and participates in propelling the car. In freeway driving, it is mostly the engine that drives the car. Still, the kinetic energy of the car is recaptured and used to charge the batteries during braking. Since the engine does not principally drive the car in most cases, and since it does not normally start the car from rest where the most torque is needed, and therefore does not need to be hugely powerful to provide large accelerations that are needed only a fraction of our driving time, the engine can be much smaller and can mostly run at a constant rate at its most efficient state. Therefore, it will have the best possible efficiency at the lowest possible weight (although manufacturers are increasing the size of engines to satisfy our hunger for more power, albeit at the cost of fuel efficiency!).

So why is it that the efficiency of a hybrid vehicle is even better in city driving than in freeway driving? Do we not accelerate, decelerate, and brake more often in city driving and therefore waste more energy? Should it not be that gas consumption in freeway driving should be even less than in city driving? Then why is it larger for hybrids?

There are two reasons for this anomaly. One is that hybrids switch to their engines in freeway driving. Therefore, the efficiency is that of the engine. However, the more important reason is drag resistance. When an object moves through the air—regardless of its shape—the faster it goes, the larger the air resistance. The very simplified equation describing this phenomenon is:

$$F_D = \frac{1}{2}\rho C_d A v^2, \tag{1.5}$$

where F_D is the drag force indicating resistance to movement, is the density ρ of the fluid (in this case, air), C_d is the coefficient of drag, a measure dependent on the shape of the object, A is the frontal area of the object (the area of the vehicle if you were to look at it directly from the front), and v^2 is the square of the velocity of the vehicle. Coefficient of drag varies for different shapes; airplanes have a much smaller coefficient of drag than buses because they are more aerodynamic. Coefficient of drag for cars also varies depending on their shape. The most important element of Equation (1.5) is speed because it is squared. Therefore, as the speed increases, the drag force increases quadratically. For example, for the same car, when everything else remains the same, as the speed goes from 30 miles an hour to 60 miles an hour, the drag increases four times as much.

At 75 miles an hour, the drag is 6.25 times as much. Therefore, in freeway driving, the engine has to provide more power to compensate for the drag force at a much higher value than city driving where speeds are low. Unlike stop-and-go driving conditions where much of the kinetic energy is converted back to electrical energy and restored into the batteries, all the energy is used to compensate for drag force and not recaptured in freeway driving, and therefore, the efficiency of hybrids in freeway driving is less than in city driving. Strange, but true.

1.7   COMMON MISCONCEPTIONS

Since we are talking about energy, let's also talk about some misconceptions people have about it.

   I. Heating a room faster by setting the thermostat higher: Some people think that if a room is very cold, they can warm it faster if they set the thermostat at a higher temperature. Unless they have variable-rate furnaces with multiple burners, variable-rate or multiple fans, and a smart controller, this is not true (most systems are simply on-off systems). Let's say the room is at 60°F and the desired temperature is 72°F. If you set the thermostat at 72°F, the heater will pump heat into the room until it gets to 72°F and stop (depending on the settings of the furnace, some systems slow down a bit when the temperature is near the desired value to prevent overshooting and to allow the furnace to cool down). Setting the thermostat at 85°F will not increase the rate of heating the room to 72°F because the furnace works at its maximum power regardless of the set temperature. So, the temperature of the room will not increase any faster; it just continues to increase until it reaches the desired temperature. So, if you initially set the thermostat to 85°F and subsequently reduce it to 72°F as it reaches this desired value, the rate of heating will be same as if you were to initially set it at 72°F and be done with it.

   II. Cooling a room by leaving the refrigerator door open: On a hot summer day, when it is difficult to bear the heat, it is tempting to leave the refrigerator door open in order to blow cool air into the room, and some people do so because they think they are generating cool air that can reduce the temperature of the room. As was discussed earlier, this is a misconception too. Although the refrigerator does blow cool air into the room, as long as the whole unit is inside the room, the total net effect is more heat. Leaving the door of a refrigerator open will actually make the room warmer, not colder. This is due to the fact that the refrigerator is colder than the room because the heat is transferred from it and added to the air that is blown over its condenser. Due to the ever-present friction and inefficiencies in every system, it takes a net positive amount of energy to do this, therefore adding more thermal energy to the room and further warming it. We will see about thermodynamic cycles, including refrigeration cycles in Chapter 4, but for now let it suffice to say that leaving your refrigerator door open does not make the room cooler.

   In fact, it is the same if an air-conditioning unit is completely enclosed inside a room. So how do air conditioners normally make a room cooler? By placing their condensers, where the hot air is blown out, outside of the room, whether a traditional AC unit, a unit installed in a window, or a unit whose evaporator is inside the room and its condenser is outside. All you are doing is moving the thermal energy from inside the room plus the work done by the system to the outside environment; the net effect is more heat. Figure 1.7a shows the evaporator of an air-conditioning system inside the room, where the thermal energy of the room is transferred to the coolant, thereby making the room cooler. Figure 1.7b shows the condenser unit of the same system outside of the room, where the thermal energy is rejected into the outside air.

(a)   (b)

Figure 1.7: Even though the evaporator part of the air conditioning unit is inside the room where it cools the air, the condenser part is outside in order to dissipate the thermal energy to the environment.

Similarly, a running fan inside a room will also make the room warmer because the energy spent by the motor is added to the room. However, since the moving air helps evaporate sweat from the body, and therefore cools it, it makes the person feel better. Nevertheless, the temperature rises as a result of running the fan.

III. The hand dryer in the restroom blows cold air when it starts: You may have noticed that when you use blown air to dry your hands in a public restroom, it feels that at the beginning, when we believe it should be hot to dry our hands quickly, the blower blows cold air. Later, when the hands are almost dry, it blows hot air. However, if you try to pass them under the blower when your hands are dry, you will notice that the air is warm from the beginning. The reason we feel the air is cold at the beginning is that when our hands are wet, the air evaporates the moisture on our hands, cooling it in the process. What our hands feel during this process is the consequence of evaporation and the transfer of heat from the hands while the moisture evaporates.

IV. Food cooks faster in boiling water if you turn up the heat: If your food is cooking in already-boiling water, turning up the heat will not increase the temperature of the water, and as long as the water remains boiling, it will not cook any faster. This is because when water or other fluids boil, the temperature remains constant at the boiling point. For water at sea level, this is 212°F or 100°C. This means that if you increase the amount of thermal energy through burning more gas or more electrical energy passing through the heating element, the additional energy will make the water boil faster, not hotter, and therefore, it will convert to steam at a faster rate, but the temperature will remain the same. At pressures lower than the air pressure at sea level (for example, if you go to Denver), the boiling temperature decreases as well, cooking the food slightly slower. So what is the right way to increase the temperature of the water and cook faster? Using a pressure cooker, in which the lid is completely sealed causing the pressure in the pot to increase. This will raise the boiling temperature, therefore cooking faster. However, you cannot remove the lid without first letting out the steam to reduce the pressure equal to the atmospheric pressure in order to taste the food or check it. Otherwise, it may blow up in your face! Each time you do this, a lot of energy is wasted too.

The same is true in steam locomotives. In order to increase the temperature of the water and generate more steam and more power, a pressure vessel is used. This increases the boiling temperature and increases the total energy that the steam carries, therefore achieving more power transmission. The downfall is that pressure vessels are heavier and more dangerous.

The same principle is also used in the design of a novel coffee cup that keeps your coffee at a constant temperature for a comparatively long time. The cup is double-walled, where the space between the two walls is filled with a chemical compound that boils at about 180°F, a desired temperature for hot coffee. If the freshly poured coffee is hotter than this temperature, the extra energy is transferred through the metal wall of the coffee cup to the fluid in between and heats it up to a boiling temperature. The remaining energy will boil the compound into steam. Since the heat capacity of boiling the compound is more than heating it, much of the initial excess heat energy of the coffee will be stored in the chemical compound in the form of steam at a constant temperature. As the coffee gets cooler, the heat energy of the compound is transferred back to the coffee, keeping it at the desired temperature for a longer time.

Enjoy your coffee without the danger of pouring it over your legs and burning them while you drive!

CHAPTER 2

Natural Frequencies

Vibrations, Hearing, Biomechanics, and Guitars

2.1 INTRODUCTION

Imagine that you are sitting at home on a fine morning when suddenly your whole house starts shaking. You look outside and realize a train or airplane has just passed by. You are at a rock concert and the drummer starts playing a bass beat so heavy you feel like your heart is actually vibrating inside your chest. Or you are stopped at an intersection and the car next to you is blasting reggae music with sub-woofers and your whole car starts to shake along with the beat. But how can an object have such a powerful effect on another object with which it does not even have physical contact?

What about the tires of a car shaking vigorously if they are not balanced and your cell phone vibrating when there is an incoming call? I know someone who claims that his back molars vibrate when he hums a D# note. He uses this vibration to tune his instruments when he does not have access to a tuner. What causes these vibrations?

The phenomenon that causes all of these is called natural frequency. In this chapter we will study natural frequency, vibrations, and many other related issues and see how to reduce unwanted vibrations and benefit from them when we need to and how they affect our everyday lives.

Imagine that you attach a small weight m to a spring (with spring constant k, see below), as shown in Figure 2.1 and hang the spring. Obviously, the weight will pull down the spring until the force in the spring equals the weight. Since it is in equilibrium, the mass will stay where it is without motion

Now imagine that you pull down the weight extending the spring and then let go; it will start to go up compressing the spring, stop, come down stretching the spring, stop again, and start to go up again, oscillating up and down repeatedly at a constant rate. How long will it continue to oscillate? This depends on a number of different factors, including the internal friction in the spring and air resistance (that converts the kinetic energy of the weight to heat, as discussed in Chapter 1), also referred to as *damping*. In the absence of any factors that will convert this energy into heat, the mass will theoretically oscillate forever, converting its potential energy P (energy stored in the spring as it is stretched or compressed) into kinetic energy K of the mass as described by:

$$K = \frac{1}{2}mV^2,$$
(2.1)

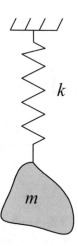

Figure 2.1: A weight hanging from a spring in a stable condition.

where V is the velocity of the mass. Of course, nothing in nature is completely frictionless, and air resistance exists unless there is an absolute vacuum. So in reality, after a few oscillations, the mass will eventually stop when all its initial energy is converted to heat.

Where did that initial energy come from? From pulling the spring down, stretching it, and storing the energy in a form called *potential energy* or *elastic energy*:

$$P = \frac{1}{2}kd^2, \tag{2.2}$$

where k is the *spring constant* and d is the displacement or the stretch in the spring from its free (unstretched) length. Spring constant is the force necessary to stretch or compress a spring one unit of length. In the metric system it is the force (in Newtons) necessary to stretch a spring one meter. In English units, it is the force necessary in lbs to stretch the spring one inch or foot. The displacement can be easily calculated from:

$$d = \frac{mg}{k}, \tag{2.3}$$

where g is the acceleration of gravity (for example, 32.2 ft/sec^2 or 9.81 m/sec^2), and therefore, mg is the weight of the mass.

It is important to notice that the little energy given to the system will cause the mass to oscillate for a long time—perhaps forever—if there is no friction or air resistance. The frequency at which the mass oscillates is called *natural frequency*.

So what is natural frequency? The natural frequency of a part or system is the frequency at which it will oscillate theoretically without any (or in reality with little) outside energy. The rate of natural frequency, in the absence of damping (such as friction) can be described as:

$$f = \frac{1}{2\pi}\sqrt{\frac{k}{m}} \text{ Hz},\qquad(2.4)$$

where the term $\frac{1}{2\pi}$ is a constant of conversion, k is the spring constant, and m is the mass of the part (not its weight). Hz (read Hertz) is the unit used to describe frequencies (for example, your radio station may be at 90.5 MHz, or 90.5 Mega Hertz of oscillations per second). Let's first explore this concept before continuing.

It should be clear from Equation (2.4) that as k increases, the natural frequency increases too. Conversely, as m increases, the natural frequency decreases. As engineers refer to this, when a system is stiffer (larger k), the natural frequency is higher and the oscillations are quick, whereas when the system is more massive (larger m), the natural frequency is lower and oscillations are slow. For example, imagine that we have two springs, one with a spring constant of 5 lb/in, one with 10 lb/in, and the weight used is 1 lb. To get the mass of this weight, we will use:

$$W = mg \quad\text{or}\quad m = \frac{W}{g} = \frac{1}{386}\frac{\text{lb}}{\text{in/sec}^2} = 0.0026 \text{ lb.sec}^2/\text{in}.\qquad(2.5)$$

(Note the unit used for mass in English system. In SI units, the unit of mass is kg, but in English units, there is no real unit for mass, so we use this unit which comes from dividing the unit of force lb by the unit of acceleration in/sec^2).

For the combination of the 5 lb/in spring and the 0.0026 lb.sec^2/in mass (1-lb weight), the natural frequency of the system will be:

$$f = \frac{1}{2\pi}\sqrt{\frac{5}{0.0026}} = 7 \text{ Hz}$$

or that one full oscillation will take $\frac{1}{7} \cong 0.14$ seconds. For the second spring, with the same mass, the natural frequency will be:

$$f = \frac{1}{2\pi}\sqrt{\frac{10}{0.0026}} = 9.87 \text{ Hz}$$

and the time required to fully oscillate once is about 0.1 seconds. As you see, when the spring is stiffer (harder, requiring more force to pull or push), the natural frequency is higher too, taking less time to complete one oscillation; the mass moves faster.

Now let's take the same 5 lb/in spring, but attach a 2-lb weight ($m = \frac{W}{g} = \frac{2}{386} = 0.0052$ lb.sec^2/in) to it. The natural frequency will be:

$$f = \frac{1}{2\pi}\sqrt{\frac{5}{0.0052}} = 4.94 \text{ Hz}$$

with the time required for a complete oscillation at 0.2 seconds.

To see how this relates to real-life products, consider sub-woofer and tweeter speakers. The sub-woofer is usually a larger, relatively heavy speaker with a more massive cone. Therefore, its natural frequency is lower, and as a result it is more appropriate for generating low-frequency bass sounds. The tweeter is usually small, such as the speaker used in your computer or cell phone, with a very light-weight diaphragm, which has a higher natural frequency that is suitable for generating higher frequency treble sounds. We will see more about these and other examples shortly.

Since systems can oscillate very easily at their natural frequency, it means that if there is a changing force that is near the natural frequency of the system, it will induce large oscillations into the system. When it is to our benefit, we take advantage of this phenomenon, whereas when it is to our detriment, we try to reduce or control it. Therefore, any time there are forces present that oscillate near the natural frequency of a system, we must watch out for large, sometimes out-of-control oscillations in the system. If the inducing force varies at a frequency that is not close to the natural frequency, the system will not oscillate freely; it requires much more energy to oscillate a system or a part at frequencies other than the natural frequency.

2.2   SYSTEM RESPONSE TO EXTERNAL FORCES AT DIFFERENT FREQUENCIES

Figure 2.2 shows the response of a system to varying-amplitude force (like a sine wave) at different frequencies. Although in reality, input forces may be very different, engineers use known inputs such as a sine wave to study the output of a system and understand its behavior. The x-axis shows the ratio of the frequency of the external force relative to the natural frequency of the system. So when this ratio is $\omega/\omega_n = 1$, the frequency of the external forcing function is the same as the natural frequency of the system. At other values, the frequency of the external force is either higher or lower than the natural frequency of the system.

The y-axis shows the response of the system, also called the *magnification factor*. It indicates how large the response of the system is relative to the amplitude of the force. So when the magnification factor is equal to 1, the amplitude of the vibration of the system is the same as the external force, and therefore, there is no magnification. When it is larger than 1, the system oscillates at a larger amplitude than the external force (it is magnified), and when it is smaller, it indicates that the vibration is reduced. This is an important factor in the design of systems where we might wish to increase or decrease the amplitude of the vibration.

Another important factor in Figure 2.2 is the damping ratio, shown as ζ (Greek symbol *zeta*). We already mentioned that all systems have some friction, air resistance, or other damping (such as shock absorbers in your car) in them. ζ indicates the level of this damping. A larger damping ratio indicates quicker conversion of the system's energy to heat, and consequently, reducing the amplitude of the vibration and how long it oscillates.

What is important about Figure 2.2 here is that it shows how the system responds as the frequency and damping ratio change. Note that around $\omega/\omega_n = 1$ (when the frequency of the external force is the same as the natural frequency), the amplitude of the response is very large,

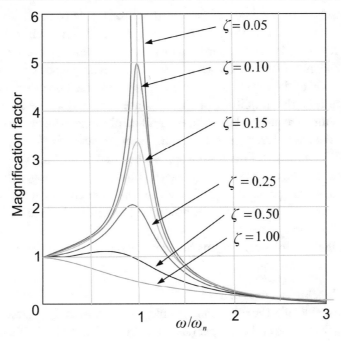

Figure 2.2: The response of a system to an external driving force at different frequencies and amplitudes and damping.

especially when the damping ratio is lower. Theoretically, in the absence of any damping, the amplitude of the response could be infinite, a theoretically devastating result. This would mean that the system could disintegrate as it is subjected to an external force at the natural frequency. However, although every system has some damping, when it is low, the amplitude of the response can be very large, many times larger than the input. As the damping increases, the amplitude of the response deceases. At the value of $\zeta = 1.00$, also called *critical damping*, the amplitude of the response varies between 1 (when the input frequency is zero) and zero as the frequency increases. This indicates that with critical damping, the amplitude of the response is always smaller than the input, and therefore, the vibration is always reduced. We will see more about this later, but as an example, the suspension system of a car is designed to have a critical damping value to prevent excessive oscillations when it encounters a bump or similar external force. If the shock absorbers (that provide the damping in your car) get old or are damaged, the car can oscillate many times before the vibration dies out. Similarly, our bodies have a lot of damping. Consequently, our body parts are largely shielded from external vibrations. Although there is still a danger present when body parts are subjected to frequencies near their natural frequencies, the damping in the body reduces this danger significantly. Next time you are taking a picture in a moving car try to place the camera somewhere on the body of the car (dashboard, side of the doors, etc.) and see how much vibration is transmitted to the camera, to the point of making it impossible to take a good picture. But when you hold the camera in your own hands, the vibration is dampened significantly, allowing you to take nice pictures. The same is true with reading a book in a moving car. Placing the book on the body of the car might make it impossible to read due to vibrations.

Also notice how the amplitude of the response increases as the frequency of the input approaches the natural frequency, but reduces significantly as the frequency of the external force increases beyond the natural frequency. For example, as the ratio of ω/ω_n approaches 2 or larger, the amplitude of the response of the system approaches zero, indicating that the system does not vibrate. It is only at around the natural frequency that the response is large. This is very important, as it indicates that we can prevent large vibrations if the frequency of the external force is larger than the natural frequency. We will later see how this plays a pivotal role in our hearing mechanism and the design of certain systems.

The following examples can help us better understand some of these concepts.

Example 2.1   Balancing the tires of a car

To make the ride of a car more comfortable, the tire assembly is attached to a suspension system consisting of a spring and a damper (also called a shock absorber). The suspension may take different forms, but in most cases it is a spring and a shock absorber. Figure 2.3 shows two typical suspension systems for cars, one with a leaf spring and shock absorber, one with a coil spring and shock absorber. The shock absorber is designed to exert a force proportional to the velocity of the oscillation in the opposite direction of the motion, therefore dampening the oscillation and stopping it; the larger the velocity of the oscillation, the larger the force will be. But as far as we are concerned, the tire-spring assembly is very similar to the system of Figure 2.1. Therefore, it has a natural frequency at which it oscillates vigorously when the frequency of the input force is close to it. Where does this force come from?

The force may come from an imbalance in the tire or the tire assembly as a result of man-ufacturing processes; no manufacturing process is ever perfect. Therefore, tires may be slightly heavier at one point compared to another, resulting in an imbalanced tire. The extra heaviness on one side of the tire, although sometimes very small (perhaps a few grams only), induces an outward force in the tire when the tire is rotating, referred to as centrifugal force. (Mechanical engineers and physicists do not refer to this name. We prefer to talk about an inward acceleration called *centripetal acceleration*, pointed toward the center of rotation. Due to the inertia of the tire, there is a reaction to this acceleration, pushing outwardly. Please see Chapters 3 and 5 for a more thorough discussion about inertia.) For example, the centrifugal force is the same force that lets you rotate a weight attached to a string without it falling, as in Figure 2.4. The same force dur-ing the spin cycle of a washing machine will force out the water. At much higher values (due to extremely rapid rotations) this force also separates uranium from other impurities, and therefore, concentrates it at higher purity levels.

Figure 2.3: Typical spring-damper assembly of automobile suspension systems.

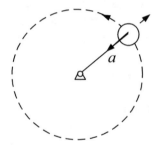

Figure 2.4: A weight, attached to a string, will not fall when it is rotating around a fixed point due to the outwardly pushing centrifugal force. Although the centripetal acceleration is inward, the reaction to this acceleration is outward.

Since this force is always outward, the direction of this force changes as the tire rotates, sometimes facing down, sometimes facing up, to the left or to the right. Especially when the force is pushing down or pushing up, it pushes against the spring, deflecting it. This is exactly the same as in Figure 2.1 where a weight (a force) is attached to a spring, causing it to oscillate. As shown in Figure 2.2, as long as the frequency of this force is not similar to the natural frequency of the suspension system and the tire assembly, the tire will not oscillate much. But when the frequency of the alternating force is close to the natural frequency of the system, even the small force caused by a few grams of the weight imbalance is enough to violently oscillate the tire, shaking the car with everything in it; it only takes a small amount of force to introduce large oscillations at the natural frequency. However, the oscillations are much smaller if the frequency of the force is below or above the natural frequency of the system.

In this case, our goal should be to eliminate this undesirable vibration. To do so, we eliminate the source of the force, the tire imbalance. This is why the tire is tested for imbalance by placing it in a machine that rotates it and measures the force exerted by the imbalance as well as its location. The technician places a counter-balance weight on the tire assembly across from the center, therefore balancing it. Since the source of the oscillating force is eliminated, so are the resulting vibrations at the natural frequency.

The same is true for any other device that rotates this way, be it the blades on the turbine of a jet engine, the tub of a washing machine when clothes are loaded in it, the driveshaft of a car, or even its engine. For example, if the blades on the turbine of a jet engine are not carefully balanced across from each other, the turbine, rotating at very high rates (perhaps 50,000 to 80,000 revolutions per minute) will generate tremendously high forces that can vigorously shake the engine. The driveshaft of your car that connects the gearbox (in the front) to the differential (placed in the rear of the car in rear-axle driven cars) must be balanced too; otherwise it will shake when rotating at the natural frequency range. And if you place a heavy jacket, a heavy article of clothing, a rug, or a blanket into a washing machine without other articles to counterbalance it, the machine may shake out of control during the spin cycle, causing it to move and break away from the water lines and cause severe water damage to its surroundings. And although it is slightly different from this exact example, even an engine has forces that must be counterbalanced to prevent excessive vibrations in it. Except for certain configurations (such as a V-6 engine that is naturally balanced), the counterbalance weights are integrated into the crankshaft.

Example 2.2   Shakers and Oscillators

In Example 2.1 we looked at a system in which our desire was to eliminate vibrations (oscillations). In many other systems we may actually want to take advantage of the oscillations at the natural frequency. Three examples (and there are many more) are a cell phone vibrator, a hand-held massager, and electric shavers. In all cases, either a mass-spring or a rotating mass system is designed with a particular natural frequency. A vibrating force, either electromagnetic or mechanical (an imbalanced-weight rotating), is applied to the system at the same frequency, whereupon the mass oscillates vigorously although the force is very small. Since little energy is needed to induce vibrations at the natural frequency rate, the cell phone vibrator, the massager, or the electric shaver head operate with little expenditure of energy.

2.3   NATURAL FREQUENCY OF OTHER COMMON SYSTEMS

Natural frequency is not just a characteristic of a mass-spring system or a rotating system. Many other systems have similar frequencies at which they oscillate with little or no external force. One such system is a pendulum. Others include wires, attached at both ends, and cantilevered beams. We will now discuss these systems because they also play an important role in many systems we often use.

2.3.1   PENDULUM

Imagine a pendulum, a mass *m* attached to a string (or bar) *l* and hung at one end as in Figure 2.5a. If you move the pendulum to one side (for accuracy, assume the angle is small) and release it, since the mass now has potential energy at an unstable state, it will move down to the bottom, gaining speed as its potential energy is converted to kinetic energy. As the mass continues to the opposite side, the kinetic energy converts back to potential energy until the mass stops, repeating the process indefinitely until, either due to friction or air resistance, the energy is lost in the form of heat. Theoretically, in the absence of friction and air resistance, the oscillation will go on forever; in reality, it will oscillate a few times before it stops. Similar to the aforementioned case with a mass and spring, if you desire to oscillate the mass at a rate other than this rate, you will have to exert a force on it, whereas at this rate, there is no need for additional input force. Similarly, this is the natural frequency at which the system oscillates with little to no external force.

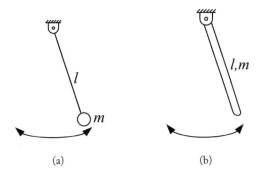

(a)   (b)

Figure 2.5: A pendulum's oscillation has a natural frequency that is independent of its mass.

Although it is easy to derive the equation describing the natural frequency, suffice it to say that the natural frequency of a simple pendulum with a point mass (not the bar) at small angles is:

$$f = \frac{1}{2\pi}\sqrt{\frac{g}{l}} \text{ Hz.} \tag{2.6}$$

Notice that the mass does not affect the natural frequency (*m* is not part of this equation). This means that regardless of the size of the mass, the natural frequency of a simple pendulum is only affected by the length of the string (or more accurately, the distance between the center of the mass and the point of suspension), and of course, the acceleration of gravity. Therefore, unless you move to a different planet, or move up a mountain, etc., the natural frequency of the pendulum will remain the same unless *l* changes.

Where is this used? An example of an application of this system in everyday life is a grandfather clock. Because the natural frequency of a pendulum is fixed, it can be designed (with appropriate dimensions and lengths) to have a period of exactly one second (or its multiples).

Therefore, at that rate, it requires very little force supplied from the spring winding, a mass hung from a chain, or an electric field, to oscillate it for a very long time. Have you ever seen how a grandfather clock is adjusted? A small screw on the bottom of the pendulum is turned to move in or out just a bit. The change in weight distribution (not the total weight) changes the location of the center of mass of the pendulum (causing a change in l) and changing the natural frequency and the period of oscillation.

The same is true in another very important part of our lives: walking and running. We will discuss this a little later together with other body parts.

Example 2.3

A child in a swing is mechanically very similar to a pendulum. Based on the size of the swing and the weight distribution of the child, the swing will have a certain natural frequency at which it tends to oscillate. It requires a lot of force to swing the child at another rate (you would have to grab the swing and move it back and forth at all times to force it to oscillate at other frequencies).

2.3.2   CANTILEVERED BEAMS

Now imagine a cantilevered bar—a bar that is attached to a rigid body (like a wall) at one end, but is free at the other end—as in Figure 2.6. If the bar is pulled down a bit and released (plucked), the bar will oscillate up and down until the stored energy in it is converted to heat due to damping or internal friction in the material or because of air resistance. Here too, if you desire oscillations other than the natural frequency rate, you will need to exert a force on the bar; it does not need any additional force to oscillate at this rate.

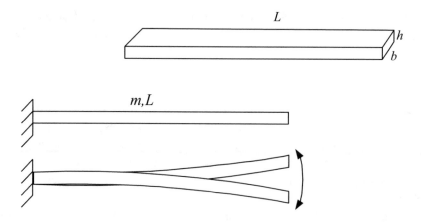

Figure 2.6: Oscillations of a cantilevered bar.

The equation describing the first natural frequency of a cantilevered beam is:

$$f_1 = \frac{1}{2\pi} \left[\frac{3.5156}{L^2} \right] \sqrt{\frac{EI}{\rho}}, \tag{2.7}$$

where L is the length of the beam, is ρ the density (or mass per unit length), and I is the area moment of inertia (see Chapter 5 for more details). The moment of inertia is discussed in Chapter 5, but for a beam with a rectangular cross section of width b and height h, as shown in Figure 2.6, it is:

$$I = \frac{1}{12} bh^3, \tag{2.8}$$

E is the modulus of elasticity, a measure of the hardness or stiffness of the material. If you pull a piece of material, it stretches. Modulus of elasticity is a representation of this relationship and describes how stiff a material is (see Chapter 5 for more detail). The Modulus of elasticity for steel is about 30×10^6 psi (in engineering terms, it is the ratio of stress over strain).

The important issue in Equations (2.7) and (2.8) is that the natural frequency of a beam is affected by the properties of the material (E and ρ), the length of the beam, its thickness and width. If any of these factors change, the natural frequency will change too. Therefore, conceivably, we can make a series of beams with different dimensions and tune them to all have different natural frequencies as we want them.

Are there any examples of where this is used? Of course there are, including the vibrations of a reed in an oboe or clarinet, a tuning fork, and our hearing mechanism. The reed of an oboe, a fundamental part of it that makes the sounds, is essentially a cantilevered bar (see Figure 2.7). As the musician forces an air-stream over it, it vibrates and generates the sound we hear after it is amplified by the body of the oboe. As the speed or pressure of the air stream changes, so does the vibration frequency of the reed. In a tuning fork too, when the legs of the u-shaped fork are struck, they vibrate at their natural frequency. Through the choice of dimensions and material used, the fork is designed to vibrate at a desired frequency.

The same principle is used in the design of a particular tachometer with no moving parts (see Figure 2.8a). Tachometers measure how fast something rotates (for example, an engine shaft). Most tachometers are based on the back-emf principle discussed in Chapters 1 and 6. The tachometer is very similar to a small electric motor that is connected to the rotating machine, and which generates a current/voltage proportional to how fast it is rotating. A gauge measures the voltage. However, this particular device has a number of small cantilevered beams of different thicknesses next to each other, each with a unique natural frequency that is slightly different from the neighboring ones. By placing the tachometer on a rotating machine, its vibrations are transferred to the tachometer, forcing only one of the beams to vibrate vigorously when its natural frequency matches that of the rotating machine. Therefore, with no moving parts, the number of revolutions per minute of the rotating machine can be measured. Figure 2.8b shows how this tachometer indicates the speed of the motor of a drill press at 2,000 rpm.

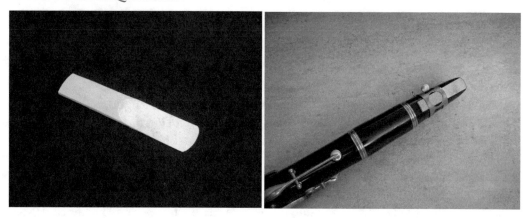

Figure 2.7: The reed of a clarinet.

2.3.3   STRINGS

Finally, consider a string that is attached to a rigid body at one end and kept taut with an axial force acting on the other, as in Figure 2.9. Plucking the string will also induce oscillations in it at its natural frequency that requires no more external force until it dies out due to internal damping and other frictional forces.

The equation [1] describing the natural frequency for the string is:

$$f = \frac{1}{2L} \sqrt{\frac{F}{\rho A}}, \tag{2.9}$$

where L is the length of the string, A is the cross sectional area, ρ is the density of the material (mass per volume, or how heavy each unit volume of the material is), and F is the tension or force in the string. When the tension is increased, the natural frequency of the string increases as well, creating a higher pitch (this is how a guitar is tuned). As the length of the string increases, the natural frequency decreases. Therefore, when the length of the string is reduced by fretting, the pitch increases (notice how the lengths of the strings in a harp are different in order to produce different pitches). For larger cross sections A, the natural frequency decreases as well. Therefore, thicker strings have a lower pitch range. Heavier materials (steel versus nylon) also produce lower pitch vibrations. Therefore, combinations of length, thickness, material, and tension can create any natural frequency we desire.

2.4   APPLICATIONS AND EXAMPLES

The following sections show the applications of natural frequencies in a multitude of systems and devices. In each case, you will notice how the same engineering principles apply and how they are

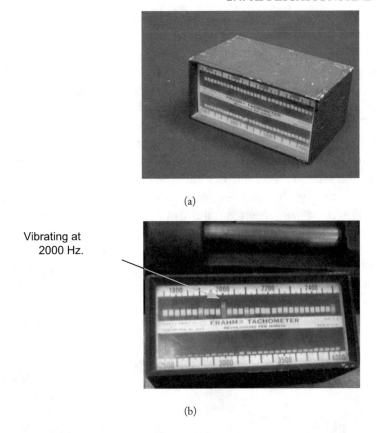

(a)

Vibrating at
2000 Hz.

(b)

Figure 2.8: A tachometer with no rotating parts.

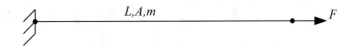

Figure 2.9: Oscillations of a string, attached firmly at both ends.

used, whether in devices and systems that we design and build, or natural systems created through natural forces.

2.4.1   GUITARS, PIANOS, AND OTHER STRINGED INSTRUMENTS

As the previous discussion indicated, large ranges of pitches can be produced by strings depending on their length, thickness, material characteristics, and tension. In a piano many strings are used, each with a specific length and thickness, practically all the same material. However, to tune the piano, an expert tuner adjusts the tension in order to create an exact pitch (natural frequency). In a piano, when a key is pressed, a hammer hits a string, and therefore, depending on how hard it hits the string, the volume of the sound varies. It is also possible to dampen the sound by pressing a damper against it. In harpsichords, the string is plucked just like a guitar; otherwise it is very similar to a piano. Harps are the same; each string at a different length produces a different pitch. The sound is adjusted by adjusting the tension. A harp is also plucked with the fingers.

In many stringed instruments, from guitars, violins, and violas to cellos and basses, all lengths are equal (some other instruments have varying lengths strings). However, the thicknesses of the strings are different and so are the materials used (Figure 2.10). Some strings are steel, some are nylon, and some are wound with a wire (nickel) for a lower pitch. The tone of the open string is adjusted/tuned by changing the tension. Subsequently, the instrument is "played" by changing the natural frequency through fretting or fingering. This is even more sophisticated in instruments such as a violin, where vibrato is common. In some electric guitars a tremolo bar is used to change the tensions of all strings simultaneously, thereby changing the pitch of all of them (see Figure 2.11). In this case, instead of attaching the strings to the body of the guitar, they are attached to the bridge. Since the bridge has a spring-loaded hinge, it can be moved slightly by the musician to change the tension.

Can you guess how musicians use a tuning fork to tune a guitar or violin? Why does it work?

Figure 2.10: Strings of a guitar produce pitches based on their lengths, the force pulling them at one end (including the changes in the force through the tremolo bar), the material from which they are made, and their cross sections.

Figure 2.11: A tremolo bar is used to change the tension on all strings simultaneously, thereby changing their natural frequency and their tone.

How Tension is Applied in String Instruments: Worm Gears: To apply tension to the strings in a string instrument either friction–based pegs or a worm gear-based tuning key is used. For a violin or a viola, where the tensions are lower and the instrument is not plucked constantly, the strings are tensioned by turning the pegs or tuning keys. These pegs are held in place through friction. The pegs and the holes are tapered at a shallow angle, and therefore, by pushing them into the hole, enough friction is generated to keep the pegs from loosening (Figure 2.12). However, in guitars and most other instruments that are plucked, the forces are larger and friction may not be enough. In that case, worm gears are usually used (Figure 2.12). So what is a worm gear? Although worm gears are not related to the subject of vibration, let's look at the way they work before continuing. This will show us how most engineering subjects are inter-related too.

Figure 2.12: In a violin, tension is provided by a tuning peg, which is held by friction. In a guitar, since the forces are larger, tension is provided by a worm gear–based tuning key.

Worm Gears: Worm gears are very common in devices for reducing speed and increasing torque, including in automobile steering mechanisms, jacks, wenches, and others. Like other pairs of gears, they provide reductions or increases in angular velocities and torques. But they also have other characteristics that make them useful in particular instances. Figure 2.13 shows a simple worm gear. Depending on whether the worm is a right-handed or left-handed worm (turning the worm in the direction of your curled fingers of the right or left-hand will move the thread forward along the direction of your thumb; also see Chapter 3), the worm gear will rotate counter-clockwise (CCW) or clockwise (CW).

Figure 2.13: A simple worm gear.

First a word about gear ratios. In all gear systems, the reduction or increase in angular speed or torque is proportional to the gear ratio (the ratio of the number of teeth on each gear, usually called the *driver gear* and the *driven gear*). Therefore:

$$\frac{\omega_1}{\omega_2} = \frac{T_2}{T_1} = \frac{N_2}{N_1},$$

(2.10)

where ω_1, T_1, N_1 and ω_2, T_2, N_2 are the angular velocity, torque, and number of teeth of each gear, as shown in Figure 2.14. Notice how these are related. The larger a gear, the slower it rotates, but the larger the torque. Therefore, by selecting the appropriate number of teeth on a pair of gears we can increase or decrease the angular speed and torque.

So why is it that when a gear rotates faster, the torque on it is lower, and vice versa? Of course we can answer this question by calculating the moments on each gear and by drawing free body diagrams as well, but here we will consider the principle of work and energy. As we have already discussed, the total energy in a system is constant unless we add energy to it or remove energy from it. This is called conservation of work and energy. Assuming that the friction in the system is small enough to be negligible, the total work or energy into and out of the system of gears is constant. However, work is equal to force multiplied by linear displacement, or equal to a torque multiplied by angular displacement. Therefore, the total input and output should be equal, or:

$$W = T_1 \times \omega_1 = T_2 \times \omega_2,$$

(2.11)

where W is the work. This is the same result as Equation (2.10). Consequently, as the angular speed is reduced, the torque is increased, and vice versa. This is exactly what happens in an automobile gear box as well. In the first gear, the output angular speed is reduced through higher gear ratios, creating larger output torques that can start a car moving. When the speed of the car increases, we shift into second and third, etc., increasing the speed, but lowering the output torque.

Figure 2.14: A gear reduction system.

Although worm gears look somewhat different, they are kinematically the same. They provide a large gear ratio for their size, but usually have more friction as well. However, depending on their helix angle, they can be self-locking, an important characteristic. So what is the helix angle? In fact, if you look closely, the worm looks like a screw. A screw is nothing more than a triangle, wrapped around a cylinder, as shown in Figure 2.15. The angle of the triangle at the tip is the helix angle. This determines how many threads are present in any given length (e.g., 20 threads per inch in a 1/4-20 screw). In reality, when a screw is rotated, the nut moves up or down on this (inclined) plane. A larger helix angle means that the nut moves faster, but it requires a larger force to move up. Imagine that the angle is large, creating a steep incline. What will happen to a box on a steep inclined plane if the force behind it is removed? As shown in Figure 2.16, in the absence of friction large enough to stop the motion, the box will slide down. With smaller helix angles the box tends to stay and not slide. If you translate this concept into a screw, and if the helix angle is small, the nut will not move down on a screw when the load is removed, causing a self-locking mechanism. If the angle is steep, it is possible that as the force is removed, the nut may automatically move down, making it not self-locking. But which one is better? Imagine you use a jack to raise your car by applying a torque to the handle. One common design for automobile jacks is equivalent to a nut moving on a screw, raising or lowering the car. How would you like it if the car you just raised would comeback down as soon as you released the handle? Here we want to make sure the nut on the jack is self-locking. In other applications such as in a hand-press we want to make sure that as soon as the handle is released the press returns without external effort by the operator. This will increase the efficiency of the system. Here, a not-self-locking screw is better. Consequently, based on our needs, we can design the screw to be self-locking or non-self-locking.

For the guitar, we obviously want the tensioner to be self locking; otherwise, as soon as the tuning keys are released the tension will be lost. Although other ways exist to do this, worm gears are commonly used because they can easily be self-locking, even at large tensions.

Figure 2.15: An inclined plane wrapped around a cylinder creates a screw.

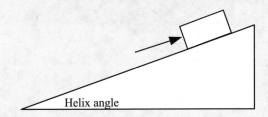

Figure 2.16: A box moving up an inclined plane.

2.4.2 SPEAKING AND VOCAL CORDS

Humans speak and produce sounds by expelling (or modulating) air from their lungs through vocal cords (also called vocal folds) situated in our larynxes. The air causes the cords (actually folds) to vibrate. As in a guitar or violin, the sound resonates within the larynx, sometimes with additional harmonics, creating an audible and unique voice. The shape of the cords, their thickness and size, and the shape of the larynx create each person's unique voice as well as the different frequencies of each sound. For example, the average fundamental frequency is about 210 Hz for women, about 125 Hz for men, and more than 300 Hz for children. By changing tension in the cords, humans can alter their frequency and produce different sounds, pronounce different letters, and sing.

The production of sound is not the only characteristic that follows the engineering principles we have already discussed. We can also see the effect of similar variables in the system. For example, generally adult males' voices are lower-pitched than those of women or children. As we might expect, the male vocal cords are longer, ranging between 1.75 and 2.5 cm (0.75 to 1 inch) versus 1.25 and 1.75 cm (0.5 to 0.75 inch) for women. We have already seen that longer strings (and cantilevered beams) have lower natural frequencies and produce lower-pitched sounds than shorter strings. As a child grows and his or her cords elongate, his or her voice changes too. Figure 2.17 shows typical vocal folds.

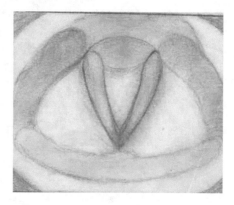

Figure 2.17: The vocal cords and folds.

2.4.3 TUNING TO A RADIO OR TV CHANNEL

The tuning of a radio or TV to different channels or stations is in fact related to natural frequencies as well, although in this case it relates to the natural frequency of an electronic circuit's output. There are certain circuits that are specifically designed so that their output voltage oscillates, e.g., like a sine wave. Although most circuits are too complicated to discuss here, we can consider a very simplified set up to study the fundamentals. This will teach us how a tuning system works. So first let's talk about this, then about tuning.

Imagine a very simple electric circuit composed of a coil or inductor L and a capacitor C as in Figure 2.18. As was discussed in Chapter 1, since the coil is a conductor, when a current passes through it, a magnetic field is developed. Conversely, when the coil is placed within a varying magnetic field, a current is induced in it. These are called electromotive force (emf) and back-emf. The important thing to realize is that these can happen as a result of each other.

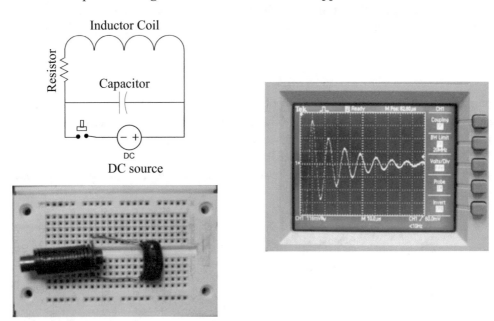

Figure 2.18: An R-L-C circuit and its response. The voltage oscillates at the natural frequency rate of the circuit.

A capacitor is another electronic element that can store electrical energy and discharge it back into the circuit when the voltage of the load is less than the voltage across the capacitor. Theoretically, if there is absolutely no loss of energy in the circuit due to electrical resistance, when the circuit is initially energized, the flow of the electrons in the circuit will cause the coil to generate a back-emf voltage which charges the capacitor. When the voltage in the coil becomes less than the capacitor, the capacitor discharges its energy back into the coil, causing the same back-emf. This repeats forever, creating an oscillating voltage. In real life, every electrical element has some resistance R, so every time the current goes through, part of the energy is converted to thermal energy, and consequently, the oscillation of the voltage in the circuit dies very quickly. However, just like a mechanical device, such as a grandfather clock where the energy loss is compensated by the energy stored in a weight or a spring, the energy stored in a battery or similar device can compensate for the lost energy in the circuit. Therefore, we can expect that the R-L-C system may continue to oscillate indefinitely as long as we have a source of external energy to compensate for the loss. In most systems a crystal is used for this purpose, but the principles stay the same. Figure 2.18 shows the schematic of an R-L-C circuit, a simple circuit consisting of a coil and a fixed capacitor put together for testing, and the output of the system as seen on an oscilloscope when an impulse signal is applied to the circuit. Notice how quickly it dies out due to the electrical resistance in the wires.

The frequency at which the voltage in the system oscillates is a function of the *capacitance* of the capacitor C (a measure of charge-storing capability of the capacitor) and the *inductance* of the coil L as:

$$f = \frac{1}{2\pi} \sqrt{\frac{1}{LC}}. \tag{2.12}$$

Changing the value of L or C will change the oscillating frequency of the circuit. This is exactly what is done in manually tuning an old-style radio by a knob. Turning the tuning knob moves a set of plates within a capacitor relative to the counterpart fixed plates, changing how much energy is stored between the plates (Figure 2.19). Although the same can be accomplished by other means (such as the use of a quartz crystal), the basic idea is to create an oscillating voltage in a circuit.

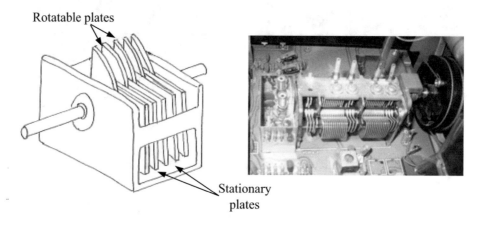

Rotatable plates

Stationary plates

Figure 2.19: A schematic drawing of a variable capacitor.

How is this related to tuning to a radio or TV station? To see this, imagine a pendulum oscillating at a particular rate, in front of which is a plane with a hole, also oscillating as in Figure 2.20. An observer is looking through the hole trying to see the pendulum. If the rate of the movements of the pendulum and the plane are exactly the same and they start at the same time, the observer will continually see the pendulum through the hole. However, if the rates are not the same, even if they start exactly at the same time, the observer will actually not see the pendulum, except by chance when they happen to be at the same location at the same time. When the two have the same frequency of oscillation, we can say that they are tuned (synchronized) with each other, moving at the same rate. Stay tuned, as we are not there yet. Now we need to see how different broadcasts are coded for distinction.

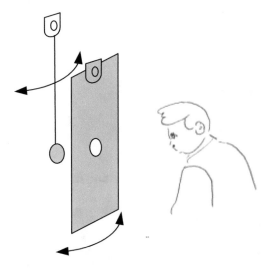

Figure 2.20: An observer behind a plane with a hole may or may not see the pendulum moving depending on whether or not the pendulum and the plane have the same frequency of motion.

There are hundreds of stations that broadcast radio and television programs. Without some unique feature to distinguish one signal (or station) from another, every receiver would capture the combined broadcasts from all the stations at once, obviously a completely useless system. This would happen if the information broadcast by any station (radio, TV, the Police, etc.) consisted of only the intended signal (for example the music) without a distinguishing signature to differentiate it from another. However, to create multiple stations with multiple channels of broadcast, each with a unique signature, the signal is modulated with a carrier signal before broadcast, either based on amplitude modulation (AM) or frequency modulation (FM). We will not get into too much detail about this, but let's see what this means.

Imagine that $f(t)$ (some function of time, which in general can be anything, including music, video, or any other signal) is the signal that is to be broadcast. Figure 2.21 shows a simple sinusoidal function $f_1(t) = \sin(t)$ as an example. Now consider another function $f_2(t) = 0.5 + 0.5\cos(50t)$ as shown Figure 2.22a (a cosine that oscillates between 0 and 1 instead). The frequency of this signal is 50 times as large as the sine function of Figure 2.21. Similarly, Figure 2.22b shows a similar signal, but at a frequency of 100 instead of 50. Notice how the two signals look the same, but one is faster at a higher oscillation frequency.

Modulating (combining, in this case multiplying) the two signals together will result in a signal that has the overall shape of the lower frequency signal $f_1(t)$, but with the higher frequen-

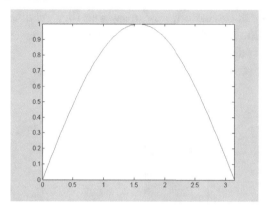

Figure 2.21: A simple sine function signal of $f(t) = \sin(t)$.

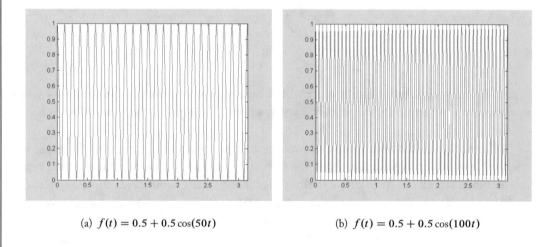

(a) $f(t) = 0.5 + 0.5\cos(50t)$ (b) $f(t) = 0.5 + 0.5\cos(100t)$

Figure 2.22: A higher frequency carrier signal at the frequency of 50 and 100 cycles per second.

cies of the second function $f_2(t)$. Figure 2.23 shows the result of modulating these functions at different frequencies of 50 and 100 as:

$$F(t) = f_1(t) \times f_2(t) = \sin(t) \times [0.5 + 0.5\cos(50t)]$$

and

$$F(t) = f_1(t) \times f_2(t) = \sin(t) \times [0.5 + 0.5\cos(100t)]\,.$$

What is interesting is that the same can be done at any other frequency, all resulting in the same overall shape of $f_1(t)$, but at different frequencies.

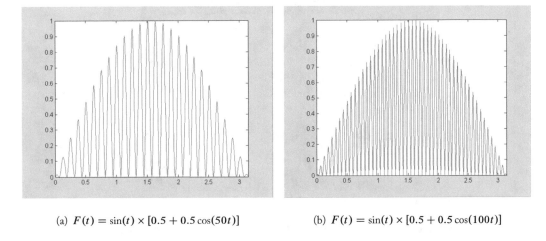

(a) $F(t) = \sin(t) \times [0.5 + 0.5 \cos(50t)]$ (b) $F(t) = \sin(t) \times [0.5 + 0.5 \cos(100t)]$

Figure 2.23: Modulated signals of a simple sine function and higher frequency cosine functions result in the same basic overall shape of the sine function, but at a higher frequency of the carrier signal.

Figures 2.24 and 2.25 show another signal and its modulated signals at two different frequencies. Similar to the previous case, the original shape of the signal is preserved but when modulated, the signal contains the higher frequencies of the carrier signals. Figure 2.26 shows two additional signals and their modulated versions for comparison.

Then how is this used as a unique signature for each broadcasting station? Imagine that each station is granted a particular frequency that it uses as its carrier frequency, used to modulate its particular signal. Whether music, dialogue, pictures and video, or any other data, the signal is modulated with the station's signature frequency. Therefore, the broadcast signal will have the basic information in it, but is broadcast with a carrier frequency unique to the station.

Now look back at Figure 2.20. As with the pendulum and the plane with a hole, where they are either in tune or out of tune, your receiver (TV, radio, or other device) may be in tune with a particular signal frequency or out of tune with it. If it is in tune with a signal, it will "see" the signal continuously and will therefore receive it. Since it is out of tune with all other signals it will not "see" any of them. All it takes for your receiver to tune in is to have the same frequency as the carrier frequency of the signal (or station). In other words, if the receiver has a frequency similar to the carrier frequency of the broadcast signal, it will receive it; if not, it will not see the broadcast signal. This is done by an oscillating circuit such as in Figure 2.18 and Equation (2.12). A variable capacitor in an *R-L-C* or similar circuit changes the natural frequency of the receiver, matching it with the frequency of the carrier signal of the particular station in which one is interested. When the two are in tune, the receiver will receive only that signal. A low-pass filter eliminates the high carrier frequency (called de-modulation), ending up with the original signal that is amplified and played back as music, dialogue, video, etc.

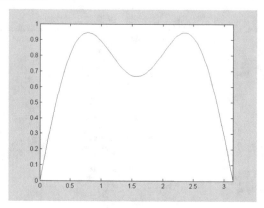

Figure 2.24: The signal $f(t) = \sin(t) + \frac{1}{3}\sin(3t)$.

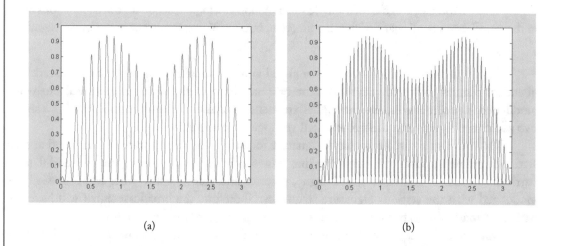

(a)   (b)

Figure 2.25: The result of modulating the signal of Figure 2.24 with a carrier signal at two different frequencies as $F(t) = [\sin(t) + \sin(3t)] \times [0.5 + 0.5\cos(50t)]$ and $F(t) = [\sin(t) + \sin(3t)] \times [0.5 + 0.5\cos(100t)]$.

Frequency modulation is somewhat more complicated. Instead of modulating the amplitude of the signal with the frequency of the carrier signal, the frequency of the carrier signal is changed based on the amplitude. As the amplitude of the signal changes, the amplitude of the carrier signal remains the same, but its frequency changes within a certain range. The rest is the same. We will not discuss the details of FM modulation here.

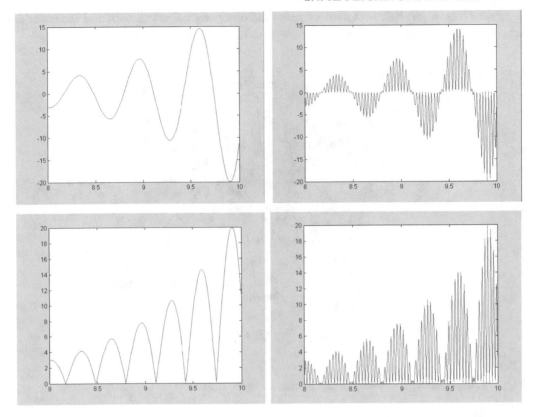

Figure 2.26: Two additional examples of signals $f(t)$ and their modulated signal $F(t) = f(t) \times [0.5 + 0.5\cos(100t)]$. Notice how the shapes of the signals are preserved at the frequency of the carrier signal.

Realistically, the modulating frequencies are very high, in the hundreds of thousands (kHz) for AM, and in the millions (MHz) for FM.

Amplitude modulation is prone to noise, but has a better range, while FM is less prone to picking up noise, but does not travel very far. This is why most available out-of-town stations where cities are not close by are AM, not FM.

2.4.4 HEARING

The hearing mechanism in humans (and most animals) is also related to natural frequencies. To see this relationship, let us first examine the human ear and its parts, then we will discuss the mechanism of hearing.

The human ear has three distinct parts: the outer ear, the middle ear, and the inner ear, as schematically shown in Figure 2.27. Each section has a different function.

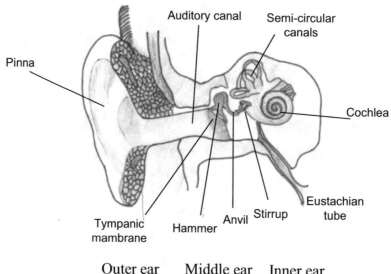

Figure 2.27: Schematic drawing of the human auditory system.

The outer ear has the following roles:

1. The *pinna* is a distinct feature of the human face, giving it a certain beauty and making the human face look as we have come to know and love.

2. It is sensitive to touch, temperature, and other stimuli.

3. It acts as a radiator to dispense of excess heat when needed. There are a lot of blood vessels in this organ. When the body needs to dispense heat, blood flow to the outer ear increases. This is why the ears turn red when the person is hot or nervous.

4. It collects sound. The sound we hear is the result of the reaction of our hearing mechanism to the vibration of molecules of air. The pinna increases the ability of the system to sense these vibrations.

5. It helps in determining the direction of the sound. Humans hear in stereo; this means they can sense the approximate location of the source of sound in space. This is because the distance from a source of sound to each ear is slightly different. The very small difference in the time that it takes for the sound to reach each ear is detected by the brain, helping it

to determine the location of the sound. However, we can also determine if the source is in front of or behind us, even with closed eyes, because of the unique shape of the outer ear. Since its shape relative to the front or rear is different, it can detect whether the source is in the front or rear.

Sound energy (a mechanical type of energy) travels through the ear canal (also called the *auditory canal*) to the *tympanic membrane* (ear drum). The tympanic membrane is a thin skin layer that is vibrated by sounds ranging in frequencies between about 20 Hz to about 20,000 Hz, an amazing range. This is why we can hear sounds within this approximate range. We cannot hear sounds with frequencies above this range (called ultrasound) or below it. Other animals can. Dogs, bats, and many other rodents can hear frequencies far above this range. Can you guess why? Primarily, it is the ability of their (smaller and therefore higher natural-frequency) tympanic membranes to oscillate at higher frequencies that enables them to hear those frequencies.

This characteristic is used to drive rodents away from houses and farms without affecting humans. Since humans do not hear ultrasonic vibrations, a device that is plugged into an electrical outlet or pushed into the ground creates loud ultrasonic sound bites annoying rodents and bats and ground squirrels and driving them away without humans even hearing it. If lower frequencies of ultrasound were used domestic animals might also hear the sound and be annoyed.

The vibrations of the tympanic membrane are transferred to the middle ear. The middle ear consists of three bones (Ossicles) called the *hammer, anvil,* and *stirrup.* These bones are held together by tiny muscles. They bridge the tympanic membrane on one side and the cochlea of the inner ear on the other, creating a physical connection between the outer ear and the inner ear. Stirrup bone touches the cochlea at the *oval window*, where the vibrations of the stirrup are transferred to the liquid within the cochlea. A narrow tube called the *Eustachian Tube* connects the *nasopharynx* (throat cavity) to the middle ear. The middle ear has four distinct functions as well:

1. It helps in isolating the inner ear from the tympanic membrane. There are many cases where the tympanic membrane may be damaged by external factors such as extreme sounds or intentional operations (e.g., when a doctor inserts a plug into the tympanic membrane to help young children with drainage of the middle ear when it is infected). If the inner ear were directly attached to the tympanic membrane, any physical damage to the tympanic membrane, whether intentional or accidental, would permanently damage the inner ear resulting in permanent hearing loss. But with this arrangement, where the middle ear acts as a safety device, damage to the outer ear will not result in a permanent loss of hearing.

2. The Eustachian tube helps with the equalization of air pressure between the outer and middle ear. Without this equalization, not only would the middle ear ache terribly, as the outside air pressure changes, the pressure difference between the outer and middle ear would prevent us from clearly hearing sounds. By swallowing, we force the Eustachian tube to open, therefore equalizing pressure in the middle ear. If an individual with a cold or flu or

allergies cannot equalize the air pressure due to inflammation of the Eustachian tube, he or she may have pain and may not hear well. Physicians may even suggest that the individual avoid flying.

3. The specific arrangement of the three bones allows the middle ear to amplify sound vibrations. This amplification helps in hearing lower threshold sounds.

4. The three bones of the middle ear transfer the vibrations to the inner ear. In the presence of very loud sounds, the vibration of these bones becomes very large too. As a safety device, and to prevent the inner ear from permanent damage, the tiny muscles connected to the hammer, anvil, and stirrup bones contract, reducing the amplitude of the vibrations, the severity of the sound, and hearing damage. If you have ever experienced a heavy feeling in your ears when exposed to loud noises (such as a loud concert or gunshot), it is because the middle ear muscles were contracted. This in itself is an indication of damage to the inner ear, even if not as severe as it might have been without this safety feature.

Consequently, sound vibrations are transferred to the inner ear.

The inner ear consists of the cochlea and the semicircular canals. We will discuss the function of the semicircular canals shortly, but they are not part of the hearing mechanism.

The cochlea is a spiral canal, about $2\frac{5}{8}$ turns, with a complicated structure whose individual functions are not yet fully understood. Within it are three passages, two of which are separated by a membrane called *basilar membrane*. Unlike the cochlea, the basilar membrane is thicker and narrower at the base of the cochlea near the oval window and thinner and wider at the apex (end). As the fluid inside the cochlea is vibrated by the ossicles (hammer, anvil, and stirrup), the basilar membrane vibrates with it. However, since the input to the cochlea is only through the oval window as one signal only, the cochlea has to decompose and codify the sound into different frequencies that can be recognized by the brain. This is the job of the basilar membrane.

Since the width and the thickness of the basilar membrane varies throughout its length, each location on it has a particular natural frequency. As the sound vibrations go through the cochlea, one location on the basilar membrane vibrates in synch with the particular frequency of the sound. It is as if each location is tuned to vibrate at one frequency, higher frequencies at the base where the basilar membrane is thicker and narrower (resulting in a higher natural frequency) and lower frequencies toward the apex where the membrane is thinner and wider (resulting in lower natural frequencies). As a result, although only one set of vibrations enters the inner ear, the basilar membrane "decomposes" the sound into individual frequencies at each location.

Along the membrane are rows of inner and outer *hair cells* (which are not really hair, but very thin cell-structures) that are extremely sensitive to motion. The outer hair cells (numbering about 12,000) help with tuning the basilar membrane for its decomposition task. The inner cells, numbering about 3,500 in a single row, detect the vibrations of the basilar membrane and send a signal to the brain through the auditory nerve, where the sound is heard and recognized (the mechanism by which the brain interprets and understands the sound and the meaning of sounds is beyond the scope of this book).

All sounds can be decomposed into a collection of simple sine and cosine sounds at particular frequencies (called the *Fourier Series*). Therefore, the collective vibrations of the individual hair-cells within the cochlea will enable us to hear and understand the sound about us. If a hair-cell does not send the proper signal, we will not hear the corresponding frequency. This is why people who lose their hearing ability will have a difficult time understanding sounds even if it is amplified by a hearing aid; they do not hear the sounds correctly.

As we age, we naturally lose our ability to hear higher frequency sounds. However, exposure to loud noises can also eventually damage hair-cells permanently, and consequently, hearing ability.

2.4.5   WALKING AND RUNNING, HEARTS AND LUNGS

Have you noticed that your arms and legs are in fact very similar to pendulums? Granted, each arm or leg has two oscillating portions, the upper part and the lower part. This is called a *double pendulum*. One fortunate thing is that the motions of both arms and legs are relatively limited; they move less than 150°. Otherwise, their motions would be more complicated. Nonetheless, each arm or leg functions as a pendulum, and like pendulums, they have a natural frequency. Equation (2.6) for Figure 2.5 is the natural frequency of a pendulum like in a grandfather's clock, with the mass concentrated at one point. The arms and the legs are more like bars with distributed mass. The natural frequency of a bar can be expressed as:

$$f = \frac{1}{2\pi} \sqrt{\frac{mgr}{I_o}}, \tag{2.13}$$

where I_0 is the second mass-moment of inertia (see Chapter 5), m is the mass of the bar, g is the acceleration of gravity, and r is the distance from the pivot point to the center of mass of the bar.

Both the length of the arm or leg (as in r) and the mass are important factors, as is the mass-moment of inertia which is a measure of the distribution of mass. There are relatively simple ways of measuring the mass of the arm or leg and calculating its mass moment of inertia even for living humans. Therefore, we should be able to calculate what we need. However, what is of interest to us is not the calculation, but understanding natural frequency and its role in our everyday life.

A person may be able to walk for hours without getting tired. But if he or she is walking briskly or carrying a weight in his or her hands while walking, even if it is only a couple of pounds, it becomes much harder to walk more than a short time before the person feels tired. Why? When walking, we tend to move our legs and arms in about their natural frequency rates. Our arms move in the opposite direction of our legs in order to help us with balance as we move forward. At these rates, it takes little energy to move the legs and arms, so a person can do it for a long time without tiring much. However, moving at a brisk rate will change this situation; now you are forcing arms and legs to move at rates different from their natural frequencies. Therefore, much more energy

Figure 2.28: The Tacoma Narrows Bridge (Prelinger Archives).

is needed to do this. If you are trying to exercise or burn calories, this is the right thing to do. If you want to walk longer, perhaps to have a nice walk along the beach, then brisk walking is not the right thing. Similarly, if you carry a small weight in your hand as you walk, it is not the weight that causes you to burn calories or get tired; it is how the weight adds to the moment of inertia and how the natural frequency of your arm changes. Therefore, even if you move at a normal rate, you still burn additional calories because at your normal rate of walking, you are no longer moving your arms at the previous natural frequency rate. Have you noticed how physical activity experts prescribe exactly the same things—to walk briskly or to carry a light weight in your hands—as you walk? This is why.

It is actually very similar for the lungs and the heart. We breathe at the natural frequency of our lungs; therefore little energy is needed to do so. But now try to breathe at a different rate, and you get tired quickly. The same is true with a heart if it beats at a rate above the natural frequency rate. By the way, dogs pant at a high rate compared to, for example, humans. Can you tell why?

An adult human heart beats at about 70 times per minute. However, for infants, it is about 120 and for young kids, it is about 90 beats per minute. Why? Since the mass of an infant's heart is smaller, its natural frequency is larger (as we have seen with other systems). As the heart grows, the rate decreases. By the way, in a 65-year lifetime, the heart beats at least $70 \times 60 \times 24 \times 365 \times 65 = 2.4 \times 10^9$ times. This is 2.4 billion times. Not too bad.

Example 2.4   Tacoma Narrows Bridge

In 1940, the brand new Tacoma Narrows Bridge in Puget Sound, Washington, collapsed due to a phenomenon called *wind-induced flutter*. The fundamental reason for the violent movements was that the girders (deep I-beams used in the construction) moved as a reaction to the high winds in the narrows because the sides of the bridge were closed and the wind could not freely move through them. Since the frequency of the variations in the wind happened to be close to the natural frequency of the bridge, the motions became larger and the bridge swayed more until it collapsed. Fortunately, since this had happened from the time of construction (but never to this extent), the bridge was closed and there was no traffic on it. A new bridge was dedicated in 1950. That bridge still stands.

2.5   BIBLIOGRAPHY

[1] Dimarogonas, Andrew: *Vibration for Engineers*, 2nd ed., Prentice-Hall, NJ, 1996. 34

CHAPTER 3

Coriolis Acceleration and its Effects

Bikes, Weather Systems, Airplanes, and Robots

3.1 INTRODUCTION

Oregon, Washington, and California on the West Coast of the U.S. are next to an ocean as are their East Coast counterparts, including New York, Florida, the Carolinas and New England states. However, their weather patterns are significantly different. For example, except for the high altitudes of the mountainous areas of California, the rest of the state does not get any snow in the winter and is relatively dry in the summer, but New York gets a lot of snow in the winter and is very humid and warm in the summer. For example, during the month of February in 2015, the eastern half of the U.S. experienced record low temperatures and record amounts of snow, even as far south as Florida, which plunged to low 20-degrees F. At the same time, the western states had a record dry season and high temperatures. The cities in the east were in single-digit temperatures even without the effects of wind-chill while California was enjoying temperatures in the 70s and 80s F.

Looking at weather maps, you could clearly see how the so called Siberian Express air mass, moving over the North Pole and traveling south over the North America veered eastward inundating the eastern part of the U.S. Why? Weather patterns are affected, among other things, by altitude, latitude, longitude, and proximity to large bodies of water, but also by a phenomenon called *Coriolis Acceleration* that pushes the air-mass, the jet-stream, and the prevailing winds toward east. In this chapter, we will learn about Coriolis acceleration and also discuss gyroscopic effects and accelerations caused by rotating frames that explain why a bicycle does not fall when riding, why you will fall if you turn the handle of a bicycle or a scooter rather than leaning right or left, and why an airplane may rotate unexpectedly when in flight.

To learn about these concepts and understand the basics, we will need to first make a few definitions and go over some introductory issues. After that, we will return to learn how Coriolis acceleration affects our everyday lives. But along the way, we will learn a lot about mechanics and how these concepts are part of our lives too.

3.2   DEFINITIONS

Since Coriolis acceleration, like all other accelerations, is a *vector*, let's start by defining vectors.

3.2.1   VECTORS

A vector is a mathematical expression possessing a magnitude, a direction (also called *line of action*), and a sense. Values that are not vectors are *scalars*. For example, a quantity such as $10 is a scalar. It has no direction; it is only a magnitude. Bags of fruit are also scalars. Time is a scalar too; it has no direction. So is the speed of travel. It only indicates how fast one moves. However, velocity is a vector. Not only does it specify a magnitude (the speed) of travel, it also indicates the direction (and sense) of travel. For example, the line of action of the velocity vector may be 30° up from horizontal. Therefore, we know in what direction the object is moving. However, this does not yet specify the sense of travel, whether it is moving away from us or getting closer to us. Therefore, we also specify the sense at which the vector acts through an arrowhead. Figure 3.1 shows a vector $\vec{P}$ with its length representing the magnitude, its direction (30° up from horizontal), and its sense (going to the right). With the same magnitude and line of action, if the sense is reversed, the object moves in the opposite direction and the location of the object will be completely different as time goes by.

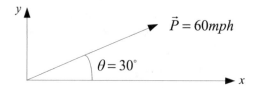

Figure 3.1: A vector with its magnitude, direction, and sense.

Force, acceleration, and distance travelled are also vectors. For example, indicating that a car moved 10 ft does not indicate in what direction it moved. To completely specify the motion, it is necessary to also specify the direction and the sense of the motion. Force is a vector too because if you pull an object it gets closer to you, and if you push the object it gets away from you. So force has a direction and a sense, and consequently, it is a vector. Notice how a vector $\vec{P}$ is specified with an arrow above it. There are also other common notations used for indicating vectors such as bold letters (**P**) or ($\underline{P}$).

Unlike scalars that can be simply added, subtracted, or multiplied, vector addition, subtraction, and multiplication require more and may yield very different results. For example, 5 apples + 3 apples = 8 apples, but a 5-lb force plus a 3-lb force may or may not be equal to 8 lbs depending on the directions and senses of the two forces. Figure 3.2a shows how a parallelogram is used to add vectors. As shown, the summation of the two vectors (also called the *resultant*) is equal to the diagonal of the parallelogram that is formed by the two vectors. The resultants of the same two vectors $\vec{V}_1$ and $\vec{V}_2$ are different when the directions and senses of the vectors change. As you can see in Figure 3.2b, adding two vectors with equal magnitudes acting on the same line of action will result in a vector twice as large if they have the same sense, but zero magnitude if they have opposite senses because they will cancel each other.

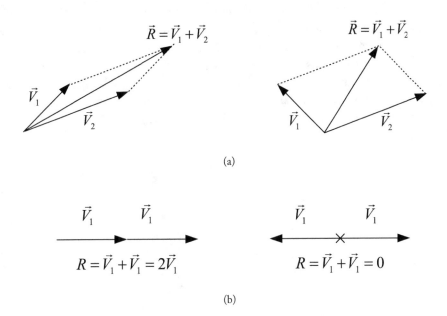

Figure 3.2: Vector additions.

Vectors are fundamentally important in engineering. Many subjects of study use vector notations and vector analysis for proper results. Examples abound, from the forces acting on the wings of an airplane in flight to the forces acting on buildings, and from hydrodynamics to space flight, motors, and robotics. For example, the force generated by a jet engine is the same as the resultant of the drag forces and lift forces that keep an airplane afloat. The forces shown in Figure 3.2a are exactly applicable to the way a ship is pulled by tugboats. The forces of the tugboats will pull the ship in the direction of the resultant force.

3.2.2   VECTOR MULTIPLICATION

Vectors can be multiplied, but not like scalars. For example, for scalars, $3 \times 4 = 12$. But for vectors, the result of multiplication is not the same. There are two types of vector multiplication called *Dot Product* ($\vec{V}_1 \cdot \vec{V}_2 = R$) and *Cross Product* ($\vec{V}_1 \times \vec{V}_2 = \vec{R}$). The result of a dot product is a scalar, a simple number. But the result of a cross product is another vector (note the difference in notations). At least for our discussion here, we need to learn about cross products.

The cross product of two vectors $\vec{V}_1$ and $\vec{V}_2$ is another vector (read it as V_1 cross V_2):

$$\vec{R} = \vec{V}_1 \times \vec{V}_2. \tag{3.1}$$

The direction of vector $\vec{R}$ is perpendicular to the plane formed by the two vectors $\vec{V}_1$ and $\vec{V}_2$, and its sense follows the *right-hand-rule*. The right-hand-rule means that if you curl the fingers of your right hand in the direction of going from $\vec{V}_1$ to $\vec{V}_2$, your thumb will indicate the sense. The right-hand-rule convention (and this is a convention only) is a very common and useful indicator, used in many different situations. Figure 3.3 shows the result of the cross product of two sample vectors. Notice the direction and the sense of the resultant vector. Also note that this is a three-dimensional or spatial figure, not planar, so you must use your imagination in seeing the vectors in three-dimensional space. Cross products will be used to explain many of the concepts related to accelerations, including Coriolis.

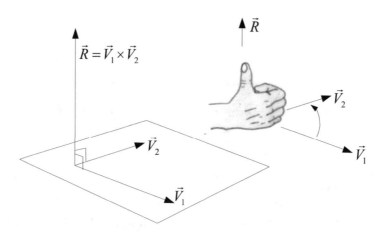

Figure 3.3: The cross product of two vectors.

For math-oriented minds, the magnitude of the dot product for simple cases is:

$$R = V_1 V_2 \cos \theta, \tag{3.2}$$

and the magnitude of the vector representing a cross product in simple cases is:

$$R = V_1 V_2 \sin \theta, \tag{3.3}$$

where V_1 and V_2 are the magnitudes of the two vectors and θ is the angle between them. One important result we can derive from Equation (3.3) is that since $\sin \theta = 0$ when $\theta = 0$, when two vectors are parallel (and therefore the angle between them is 0), their cross product will be 0. Similarly, when two vectors are perpendicular to each other, their dot product is zero.

In practice, both dot and cross products are important and very useful. For example, imagine that you push a box with force $\vec{F}$ for a distance $\vec{d}$ as in Figure 3.4. As mentioned earlier, both force and distance (we usually refer to distance as displacement) are vectors; they have a magnitude, but also a direction and a sense. The energy required to move the object is called *work*. Work, like energy, is a scalar; it has a magnitude but no direction. Obviously, the larger the force or distance, the more energy is required to move the object. To calculate the work needed to move the object, we can take the dot product of the two vectors $\vec{F}$ and $\vec{d}$, which yields a scalar, as expected. Therefore:

$$W = \vec{d} \cdot \vec{F} = dF \cos \theta.$$

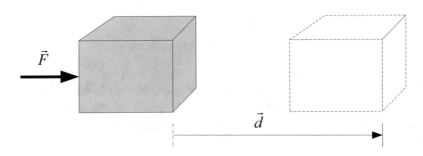

Figure 3.4: The work or energy needed to move an object is the dot product of the force and the displacement.

As mentioned earlier, when two vectors are perpendicular to each other, their dot product is zero. The weight of the box of Figure 3.4 is a vector that is directed downward (due to gravity). As the box is pushed to the right by force $\vec{F}$, the weight does not do any work because it is perpendicular to the displacement; it only does work when the box moves in the same direction, downward (we refer to this as a change in *potential energy*).

Now imagine that you are tightening a bolt using a wrench as in Figure 3.5. Obviously, if you exert a larger force or if you use a longer wrench, increasing the distance of the force to the bolt, the bolt tightens more. The cross product of these two vectors $\vec{F}$ and $\vec{d}$ is another vector called *moment* or *torque*. This torque is what tightens the bolt, and is perpendicular to both vectors, and its magnitude is:

$$\left| \vec{T} \right| = \vec{d} \times \vec{F} = dF \sin \theta.$$

So, although in both cases, the vectors involved are $\vec{F}$ and $\vec{d}$, the result of multiplying them together can be drastically different.

Can you figure out the direction of the torque vector? For Figure 3.5, it is into the page (as if an arrow is shot into the page). Notice how your thumb, with the curled fingers of your right hand going from $\vec{d}$ to $\vec{F}$, points into the page. Most common bolts have right-hand threads

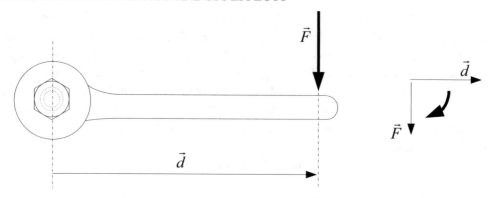

Figure 3.5: The torque caused by a force $\vec{F}$ applied at a distance $\vec{d}$.

(when you rotate the bolt or a nut in the direction of your curled right-hand fingers, the bolt or the nut moves forward in the direction of your thumb). Therefore, this torque moves the bolt forward, thus tightening it. A left-hand threaded bolt will move backward. Consequently, turning it in the same direction will loosen the bolt.

3.2.3   ROTATIONS

Now let's see how a rotational motion is defined. If you imagine a plate rotating, the direction of the rotation can be specified as clockwise (CW) or counterclockwise (CCW), depending on which side you look at (see Figure 3.6). If you are not familiar with these terms, it is probably because you have always had a digital clock. Nonetheless, we can also specify the rotational motion of the plate as a vector perpendicular to the motion.

The vector is conventionally assumed to be directed in the direction of your thumb if your right-hand fingers curl in the direction of rotation (right-hand-rule). Therefore, looking straight at a clock and curling your fingers in the direction of rotation (CW) will point your right thumb in the direction toward the clock. The opposite rotation (CCW) will point your right thumb outward away from the clock.

Looking at a bicycle too, when the tires rotate, the rotation of each one can be described by a vector as shown in Figure 3.7. This vector (and its direction) is very important in understanding why we can ride a bicycle without falling. Similarly, the rotations of the propellers of an airplane or the turbine of a jet engine can be characterized by vectors in the same fashion.

3.2.4   ACCELERATION

Acceleration is the change in either the magnitude of velocity, its direction, or both. For example, if you are currently going at a rate of 50 mph traveling south on a straight road, but your speed increases to 60 mph, you have a positive acceleration in the same direction as your travel, as

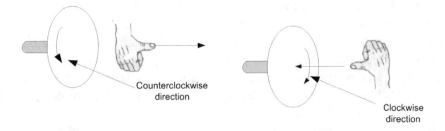

Figure 3.6: Rotation of a plate in clockwise or counterclockwise directions.

Figure 3.7: A bicycle's tire rotations can be specified by a vector.

indicated by the fact that your body is pushed back in the opposite direction of the acceleration. Similarly, if you continue to travel at 50 mph but follow a turn in the road and change your direction, you experience an acceleration as indicated by the fact that your body moves in the opposite direction of the change (we will discuss the reaction of the body to acceleration later).

If your speed decreases from 50 mph to 40 mph, it creates an acceleration in the opposite direction of your travel, slowing you down. Strictly speaking, this is not a negative acceleration, but is called *deceleration*. And in fact, there is a difference between negative acceleration and deceleration. Deceleration means you are slowing down. Negative acceleration means that the acceleration vector is in the negative direction relative to a reference frame (or positive axis). In other words, if the positive direction of an axis is to the right, and if the direction of the acceleration is in the negative direction (to the left), then this vector is negative. It may still be an acceleration (increasing the velocity to the left) or a deceleration (slowing down while still going to the left). Deceleration is in the opposite direction of your velocity or direction of motion, and therefore, it slows you down. Consequently, if your motion is in the negative direction and you decelerate (slow down), your acceleration is in the opposite direction of motion (which is the positive direction) and therefore, a positive acceleration. You should always look into the direction of the acceleration vector in relation to a reference frame to decide if it is positive or negative acceleration, as compared to whether your speed is increasing or decreasing to decide if it is an acceleration or deceleration.

There are many different types of acceleration. For example, consider a point on a rotating plane as shown in Figure 3.8. At the instant shown, the point P travels exactly to the left, and therefore, its velocity is also pointed to the left. However, from experience we know that a little later it will end up at point P', traveling down, with its velocity pointed down. Obviously, the direction of the velocity between these two points changes, even if the magnitude remains the same. Therefore, in addition to possibly changing in magnitude, the direction of the velocity vector has changed. This must have been caused by an acceleration too. This is called *centripetal* acceleration and is a function of the square of the angular velocity of the plate ω^2. As we will discuss shortly, Coriolis acceleration is another type of acceleration, and together with all other accelerations that may exist, constitutes the total acceleration of the object of interest.

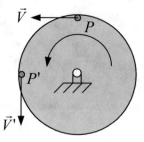

Figure 3.8: The rotation of a plate and how the direction of the velocity of any point on it changes as it rotates, causing centripetal acceleration.

3.2.5   REFERENCE FRAMES

Reference frames are used to describe the position, orientation, and motions of objects in a plane or in space as depicted in Figure 3.9. In two dimensions (on a plane), we usually use two axes, x and y. In three dimensions (space) we use three axes x, y, and z. For example, when in a plane, point P is at a distance of a from the x-axis and b from the y-axis. In three-dimensions (space) a point Q is expressed by three values of a distance a from the x-axis and b from the y-axis for the projection of Q on the x-y plane, and c from the x-y plane. Similarly, we can define the orientation of an object relative to these axes. Motions can also be defined relative to these axes within the reference frame. The axes of frames are always mutually perpendicular to each other, and they follow the same right-hand-rule we saw earlier. Therefore, the z-axis will be in the direction of your thumb if you curl your fingers in the direction of going from the x-axis to the y-axis, or $\vec{x} \times \vec{y} = \vec{z}$.

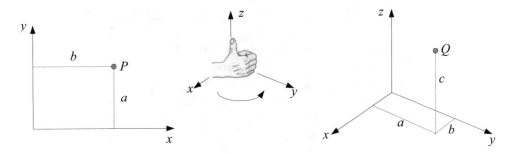

Figure 3.9: Two- and three-dimensional reference frames.

We can also consider an extension to this, which helps us with the next sections as well. The two reference frames shown in Figure 3.9 are stationary; they are fixed and do not move, and therefore we refer to other things relative to them. However, it is possible to also have additional frames that move relative to the fixed reference frame. We call them *moving frames*. For example, imagine that we attach a frame $x' - y' - z'$ to a bike at the hub of the front tire as shown in Figure 3.10. As the bike moves, the location of the frame relative to the fixed frame $x - y - z$ will change. However, unlike the location of the rider relative to $x' - y' - z'$ which does not change, the position of any point P on the tire (for example, the valve stem) relative to $x' - y' - z'$ does change. This distinction will play an important role in the next section.

3.2.6 ROTATING FRAMES

Figure 3.11 shows a wheel that is rotating about a shaft. As we discussed in Section 3.2.3, the rotation of the wheel can be described by a vector perpendicular to it. In this figure, a fixed reference frame $x - y - z$ is attached to the center of the wheel (the z-axis is perpendicular to the plane of the wheel, indicating its direction of rotation) while $x' - y' - z'$ is a frame attached to the wheel at point P. If you notice, the $x' - y' - z'$ frame does *not* stay at one location when the wheel rotates; the frame rotates with the wheel. This is called a *rotating frame*.

Looking at the same wheel from above, we will see both the fixed frame and the rotating frame. When the wheel rotates, the rotating frame moves to new locations and its position and orientation (the directions of its axes) change. Figure 3.12 shows the wheel from above. You can also see how the same applies to a rotating bar. When the bar rotates, the frame $x' - y' - z'$ attached to it rotates with the bar.

In fact, the same applies to the rotation of planet Earth and everything that moves with it. We can attach a fixed frame $x - y$ to the center of the Earth as well as frames $x' - y'$ elsewhere

Figure 3.10: A moving frame and its motion relative to a fixed frame.

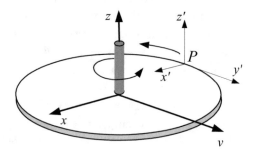

Figure 3.11: A rotating frame.

(Figure 3.13). As the Earth rotates, the frames attached to it also rotate. This rotation is very slow, one revolution every 24 hours. However, since the average radius of the Earth is 3,960 miles (6,370 km), the speed of a point on the equator is $2\pi(3960)/24$ or over 1,000 miles per hour ($2\pi(6370)/24$ or over 1,600 km/hr). Therefore, although the frame rotates slowly, its position changes vastly. Nonetheless, the frame is rotating and this does matter when we talk about the weather.

Let's take this one step further as it will shortly help us with our analysis. Suppose that a frame $x_1 - y_1 - z_1$ is attached to a rotating bar, rotating relative to a fixed frame $x - y - z$. Now also assume that a second rotating bar is attached to the first bar, and a frame $x_2 - y_2 - z_2$ is attached to this bar, as shown in Figure 3.14a. When both bars rotate relative to each other, not only do these frames rotate relative to the fixed frame, the second frame rotates relative to

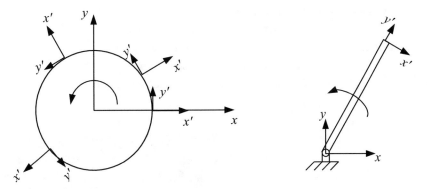

Figure 3.12: Rotating frames attached to a wheel or a bar. As the wheel or the bar rotate, the position and orientation of the rotating frame changes.

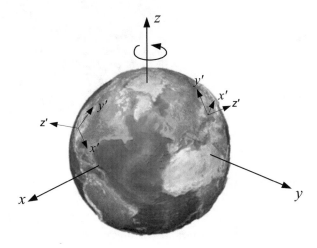

Figure 3.13: The Earth and a frame attached to it. The frame rotates with the rotation of the Earth.

the first one. The way we look at this in mechanics is to assume that you are located on the first bar (let's say there is a chair attached to this bar and you are sitting on it). Then you may *not* feel that frame $x_1 - y_1 - z_1$ attached to your chair is rotating, but you will see that the second frame $x_2 - y_2 - z_2$ is rotating relative to you. Therefore, there is motion between these frames relative to each other. In fact, this is what happens to us on Earth. Since we are attached to the Earth, we do not necessarily feel that we are rotating, but we see other objects (frames) move relative to us. However, an observer outside of planet Earth (e.g., in a spaceship or a satellite) will see us rotating and other objects moving relative to us. The same is true if you are sitting in an airplane and someone walks in the aisle. Regardless of whether or not you feel the motion of the airplane (which is moving very fast), you see the person is getting closer to you. However, you both move relative to an outside frame (or object). Please note that although in Figure 3.14 both arms move in the same plane, generally they may move in three-dimensional motion. Figure 3.14b shows a robot with its linkages moving relative to each other in three dimensions.

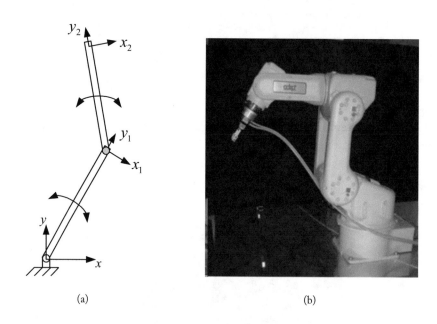

(a)   (b)

Figure 3.14: Motions of moving frames relative to a fixed frame and relative to each other.

Neither motions nor frames representing them have to necessarily be rotational. A movement void of any rotation is called a *translation*. For example, Figure 3.15 shows two simple examples where in (a), a slider simply slides (translates) on a bar while a second bar, attached to it, rotates. In case (b) the bar rotates while a slider slides over the bar. In contrast with the two-bar system of Figure 3.14 where two bars rotate relative to each other, in this case a slider translates while the bar rotates. Once again, this will be an important issue when we talk about Coriolis acceleration and the weather.

In Figure 3.15 frames are attached to the bars and sliders. In case (a), frame $x_1 - y_1 - z_1$ (the z-axis is perpendicular to the page, but not shown) is attached to the slider and translates with it while $x_2 - y_2 - z_2$, attached to the bar, rotates relative to it. In case (b), frame $x_1 - y_1 - z_1$ is attached to the bar and rotates with it while frame $x_2 - y_2 - z_2$, attached to the slider, rotates with the bar but also slides (translates) with the slider relative to $x_1 - y_1 - z_1$.

Although these two cases seem similar to each other, they are in fact very different. In case (a), the first frame $x_1 - y_1 - z_1$ only translates, while in (b), frame $x_1 - y_1 - z_1$ rotates.

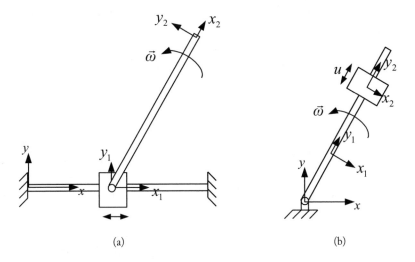

Figure 3.15: Combined rotations and translation with a bar and a slider. Although these two systems have similar components, the motions are very different.

It is therefore a *rotating frame*. Consequently, frame $x_2 - y_2 - z_2$ also rotates with it. So, it is important to note that when a frame is rotating, and within it there are other motions (either translations or rotations), the rotation of the first frame will rotate the subsequent frames. This change in the direction of the velocity of the subsequent frames creates an acceleration component called *Coriolis acceleration* that does not exist when the frame does not rotate. Now that we have endured a long set of introductions, we are ready to look at this acceleration and see what it does.

3.3   CORIOLIS ACCELERATION

Coriolis acceleration is one of the components of the total acceleration of a particle or a rigid body. The total acceleration is the vector addition of all the changes in the magnitude and the direction of the velocity of the object, each caused by something different. However, the Coriolis acceleration is present when *there is a velocity within a rotating frame*. Therefore, the first requirement is that there must be a rotating frame, within which there is another motion. These two requirements must be present for the Coriolis to exist, and it is the result of the changes in the direction of the velocity of the second motion caused by the rotating frame. Otherwise, if the frame is not rotating, or if there is no motion present on the rotating frame (therefore no velocity), there will not be a Coriolis acceleration present.

Looking once again at Figure 3.15a, notice that the first frame is not rotating, and consequently, even though the bar is moving relative to it, there is no Coriolis acceleration, whereas in

Figure 3.15b since there is a rotating frame upon which there is a velocity $\vec{u}$, there will be Coriolis acceleration.

Coriolis acceleration, like all other components of acceleration of a body, is a vector with magnitude and direction. For mathematically oriented minds, the magnitude of the Coriolis acceleration is:

$$\vec{a}_c = 2\vec{\omega} \times \vec{u}, \tag{3.4}$$

where $\vec{a}_c$ is the Coriolis acceleration component, $\vec{\omega}$ is the angular velocity vector of the rotating frame, and $\vec{u}$ is the linear velocity of the frame that moves relative to it. As you may remember, $\vec{\omega}$ is the representation of the rotation of an object, and it follows the right-hand-rule; if you curl the fingers of your right hand in the direction of rotation, your thumb will show the direction of the vector representing the rotation. Similarly, $\vec{u}$ is the vector showing the direction of the linear (translational) motion of the slider. As we discussed in Section 3.2.1, the cross product is another vector whose direction also follows the right-hand-rule. Therefore, if the fingers of your right hand are curled in the direction of going from vector $\vec{\omega}$ to vector $\vec{u}$, your thumb will be in the direction of the Coriolis acceleration. These vectors, shown in Figure 3.15b, are drawn again in Figure 3.16 with the corresponding directions of Coriolis acceleration when the direction of vector $\vec{\omega}$ changes. Notice how the direction of Coriolis acceleration changes with a change in the direction of $\vec{\omega}$. The same will happen if the direction of $\vec{u}$ changes. We will shortly see how this is an important factor in weather systems.

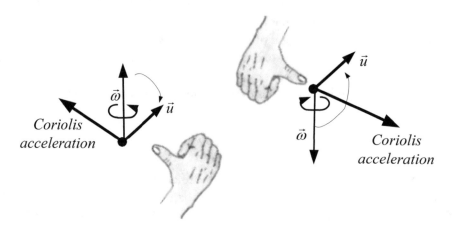

Figure 3.16: The direction of Coriolis acceleration.

3.4   INERTIAL REACTION TO ACCELERATION

Imagine you are sitting in a car and the driver presses on the gas. What happens to your body as the car accelerates? You probably have noticed that as the car accelerates forward, your body is pushed back against the seat. Similarly, if the driver brakes, creating a deceleration (or a negative acceleration, pointing in the opposite direction), your body will be thrown forward. This is why when a car is in a head-on collision the passengers are thrown forward, and if not restrained by either a seat belt or airbag, they can collide with the windshield and be severely injured.

Why is it like this? Because as we will see in Chapter 5, the inertia of a body (its mass) tends to resist changes in movement; it does not want to go faster or to slow down. If it is not moving, it tends to stay that way. If it is moving at a particular speed, it tends to continue at that speed. The only reason it might be forced to move or change its speed is if a force or torque is applied to it. In that case, the body reacts to the acceleration in the opposite direction of the acceleration. If it is pushed forward, it tends to want to move backward. If is it pushed backward, it tends to want to go forward, resisting change in its motion.

In mechanics, we do not really write our equations like this; instead, we draw what is called a free-body diagram and a mass-acceleration (or inertial-reaction) diagram, and set them equal to each other in order to solve the problem. But this is beyond the scope of this book. We will just look at the reaction of the mass or inertia, which is to resist motion, always trying to move in the opposite direction of an acceleration. We can also see the same phenomenon when a mass, attached to a string, is rotated. In rotational motions, the so-called centripetal acceleration is always pointed toward the center of rotation. Therefore, the mass always tends to move away from the center (this is referred to as centrifugal force). If it were not for the string applying a force to it, the mass would fly away from the center. Therefore, the mass continues in a circle as long as the string applies a force to it. You may realize that this is also the same as what happens to fabrics in a washing machine during the spin cycle: water particles, free to move without much restraint, tend to move away from the center of rotation while the fabric is spun fast, separating from the fabric outwardly. The same is also true in other centrifugal devices that separate solid particles from a mixture, including blood samples and Uranium concentrators used for enrichment.

In fact, this can also be applied to societies. Since each society (a family, community, city, country) has a "mass," it usually resists change unless something forces it, and even at that, the society reacts to it due to its inertia. Some societies have larger inertia (regardless of their size), some smaller inertia. The larger the inertia of a society is, the larger the resistance to change. More traditional societies have larger inertia; they do not like to change their traditions as much because of the value they see in it. In many cases, in order for a society to accept changes some large force (influence) is needed. These may be economic forces (such as periods of growth or depression), natural disasters, wars, great leaders, dictatorships, or huge social effects. Similarly, small changes in society can be accomplished with more ease because reaction to change will be smaller. In some instances, when a great change is forced on a society it may result in disastrous reactions, break-downs of the fabric of society, or revolutions.

We will see the same phenomenon as we discuss weather systems shortly.

3.5 AIR AND WATER CIRCULATIONS (CONVECTIONS) DUE TO HEAT

Before we embark on learning about the relationship between Coriolis acceleration and the motions of air masses, it is necessary to also understand one other phenomenon: the circulation of air and water due to heat injection (have you noticed how many different issues are at work for this phenomenon?).

Imagine a pot of water on a stove. As the pot heats up, the water warms up as well. But the hottest point on the pot is where it is the closest to the source of heat. At that location, the water adjacent to the hottest point receives maximum heat. As the molecules of water warm up, due to the increased kinetic energy, the distance between molecules of water increases, increasing the volume and decreasing the density of water (making it lighter) adjacent to the hottest point. The lighter water now rises up toward the surface (as for example, a piece of wood might do when placed in water—since its density is lower than water, it floats up). However, rising water cannot leave a vacuum behind; something else has to replace it. Consequently, colder water from the surrounding area will rush in to replace it. As shown in Figure 3.17, water rises near the heat source and colder water rushes in to replace it, creating a double (actually donut-shaped) circulation all around.

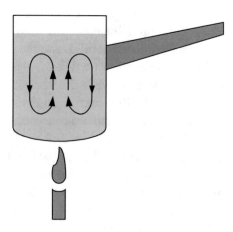

Figure 3.17: Circulation of water in a pot.

The same is true with air; as warmer but lighter air rises, colder air from elsewhere rushes in to replace it. Otherwise, we end up with vacuum (and the possibility that people in a warm place may not have enough air to breathe). This is the root cause of wind, and why in general, wind is cooler. This is also why beaches are generally windy. When the earth warms up due to sunshine, air rises. The cooler air from the ocean blows in to replace it. If you stand next to an open flame like a BBQ pit, the warm air rises and colder air replaces it, and consequently, although you may feel warmer on your frontside where the radiation heat from the fire warms you, you may feel relatively colder on your back due to the wind.

> Sometimes firefighters try to control fire by fire. This means that to combat an advancing brush fire, they start a new fire at a safe distance in advance of the burning fire. The air above the original fire, being hot, rises, pulling the air from around for replacement. This will cause the new fire, set by the firefighters, to move in the direction of the old fire until they merge. Since the brush is already burned by the new fire, the original one runs out of fuel and dies out. This is possible only because of the direction of the wind.

The faster the air moves, the lower its pressure (this is why airplanes can fly, as we will see later). Therefore, the warmer air that rises will create a lower pressure region. The region with colder air that is not moving has higher pressure (which in meteorology is considered stable air). The air from the higher pressure region flows toward the lower pressure region, creating wind.

> Most materials expand as they are heated. This is due to the increase in the kinetic energy of their molecules, resulting in increased distance between them. Therefore, the density of these materials decreases as they are heated. However, there are certain materials that do not abide by this rule. For example, Bismuth, a naturally occurring element with atomic number 83, gets smaller when heated, and therefore, more dense.
>
> Water is the same. Water expands when heated and contracts when cooled until about 4°C. At this temperature, it is at its smallest volume, and therefore, its densest. As it is cooled further and freezes, it actually expands. This expansion has many important consequences:
>
> 1. Due to this expansion when water freezes, the volume increases. This means that as water freezes it requires more space. If there is no room for this expansion, the resulting forces can be very large. For example, a bottle which is relatively full of water and is capped well may explode if left in a freezer. Similarly, exposed piping in cold environments can burst when water freezes.
>
> 2. Since water expands when frozen, it becomes lighter. Consequently, when making ice cubes, you may notice that the center of the ice cube rises as it freezes and becomes lighter and lighter.

3. Water is densest at 4°C. This means that the water in large bodies of water like pools or lakes is densest just near freezing, but not yet frozen. Because it is the densest, the water at this temperature sinks to the bottom. This keeps the fish and other living creatures safe. Otherwise, if ice were denser, it would sink down to the bottom, freezing and killing all its fish and other living organisms. Is it not nice that Nature thinks of these things?

The same happens with air surrounding the Earth at the macro level. As the air warms under the influence of the sun, it rises, pulling in colder air to replace it. However, wind patterns change as the Earth rotates around the sun, influenced by its tilt.

Earth has a tilt of approximately 23.45° relative to its elliptical path (see Figure 3.18). Due to this tilt, the total amount of sunshine received at each location changes during the year, causing the seasons. The Equator divides the Earth into two equal halves. The Tropic of Cancer is 23.45° above the Equator, and the Tropic of Capricorn is 23.45° below it. It is on these two tropics that the sun's energy is the greatest in the summer for the Northern Hemisphere and winter (their summer, even though it is January and February) in the Southern Hemisphere because each is perpendicular to the radiated energy from the sun. Many of the deserts of the world are also located on these two tropics. For this reason, wind directions change in different seasons, affecting Coriolis acceleration.

Imagine a summer day in the Northern Hemisphere; the most intense heat radiation from the sun over the Earth occurs around the lower 1/3 of the Northern Hemisphere, both on land and on bodies of water. Very similar to the pot of water on a stove, as the air and water warm up, the air and the moisture within it rise, creating a low pressure area, causing the air from the high pressure area to rush in to replace it. This creates an almost constant circulation of air throughout the Earth with circulating winds (called *cells*). However, due to other influences, instead of a double circulation pattern, there are three circulatory patterns over the Northern Hemisphere and three over the Southern Hemisphere (see Figure 3.19). These are called Hadley, Ferrel, and Polar cells. Notice that the winds go in opposite directions near the surface versus the upper atmosphere.

The expectation would be that there should only be one cell in each hemisphere; hot air rises, and is replaced by cold air. However, both Poles are huge heat sinks; they are very cold, with cold air that sinks down, creating Polar cells. In general, the weather between 0 and 30 degrees is heavily influenced by the relatively stable Hadley cell, as is the weather between 60 and 90 degrees by the relatively stable Polar cells. It is the areas between 30–60 degrees, influenced by the Ferrel cells, that are more unstable.

The Ferrel cells are somewhat secondary in nature, existing as a result of the Hadley and Polar cells. As a result, the Ferrel cells, also known as the *Zone of Mixing*, change much more than the other cells. Notice that most of the land mass in the Northern Hemisphere is in this region,

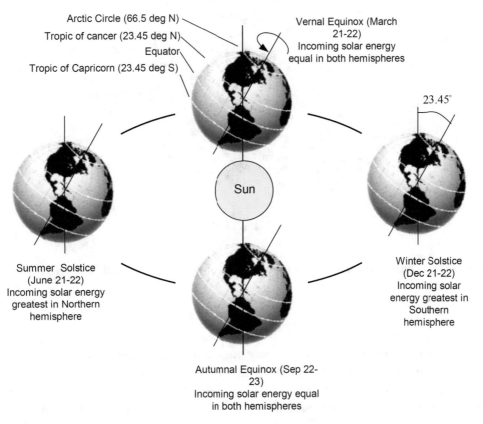

Figure 3.18: The position of Earth relative to the Sun during the year.

including the U.S., the southern half of Canada, most of Europe, most of Asia, including China, and the southern half of Russia.

3.6   CORIOLIS ACCELERATION AND WEATHER SYSTEMS

Coriolis acceleration affects many other systems too, but since we have studied so much to get to this point, let's talk about the effects of Coriolis acceleration on weather systems.

So, once again, why is it that the states along the West and East Coast of the U.S. are all adjacent to large bodies of water (oceans), but their weather patterns are so different? One would expect that being next to an ocean, the weather is heavily influenced by moisture, and therefore, all these states should have similar weather patterns, both in the summer and in winter. However, we know they do not have similar weather patterns. North east states receive significant snow during winter season while south east states do not; those states are generally warmer. Similarly,

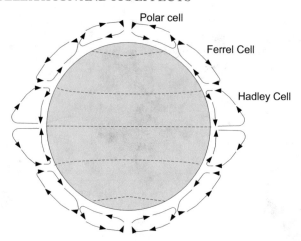

Figure 3.19: The continuous air circulation patterns of Hadley, Ferrel, and Polar cells in the Northern and Southern Hemispheres.

north west states receive more snow (although not as much as the eastern states) than the south western states. Summers in Florida are very humid, but not in California. As you see, it is not simply the latitude or longitude that causes differences in weather patterns.

Additionally, there are somewhat constant world-wide winds (the jet stream, prevailing winds, trade winds, etc.) that almost always blow in the same direction, and although they do dip down or move up as the weather and seasons change, they predominantly blow in the same direction (which has been used for sailing for millennia). What causes these winds? And what causes cyclones? Coriolis acceleration.

As we saw in Section 3.3, when there is (linear) motion within a rotating frame, there is a component of acceleration called Coriolis, which is perpendicular to both the rotation vector and the motion (velocity) vector, and it follows the right-hand-rule. Here, we have the perfect recipe for this too, where the rotating frame is the Earth, and the motions are of the aforementioned cells. The combination of the two together causes Coriolis acceleration. However, notice that since the winds are in different directions, the directions of the Coriolis accelerations vary too.

It should be clear that the wind directions at the surface and at higher altitudes are op-posites, and therefore, the directions of Coriolis accelerations will also be opposite. But to avoid confusion, let's only look at these directions on the surface; the opposite will be true for higher elevations.

First let's look at the Polar cell in the Northern Hemisphere. Figure 3.20 is a closer look at the Polar cell. The rotation of the Earth is represented as an upward vector, and the velocity vector of the air moving within the Polar cell near the surface is southward, moving away from the Pole. Remembering the Coriolis equation of $\vec{a}_c = 2\vec{\omega} \times \vec{u}$, the Coriolis acceleration will be toward the east. However, just like the reaction of your body to a forward acceleration due to its inertia (see Section 3.4), which pushes your body backward, the air mass will react to this acceleration, causing it to move toward the west, creating *Polar Easterly* winds (coming from the east). The air mass, instead of simply moving down from the high pressure area to the low pressure area, generally moves west in the 60–90 degree region.

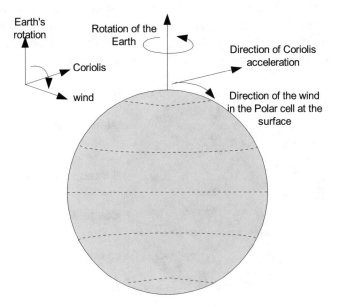

Figure 3.20: Within the Polar cell, the direction of the winds near the surface is southward. The Coriolis acceleration is toward the east, forcing the winds to react and go westward, creating Polar Easterly winds.

The rest is similar. If you look at the Ferrel cell, the motion near the surface is northward toward the Pole. Since the rotation vector is still the same, the direction of Coriolis will be toward the west and the reaction of the air mass will be toward the east, creating the Westerly winds within the 30–60 degrees region. Once again, instead of simply moving from high to low pressure areas, the winds shift toward the east. What is interesting is that as the Westerlies and the Polar Easterlies run into each other, they create all sorts of weather patterns, affecting regional climates. Westerlies bring warm and moist air from the oceans to the west coasts of the continents, describing why the weather on the East Coast of the U.S. is so different from the West Coast. Due to these Westerlies, the general pattern of air mass movements in the U.S. is eastward; air masses move from the west to the east. West Coast weather is influenced by the air coming from the Atlantic Ocean, a marine climate that is warmer and moister in winter, not causing snow until higher altitudes, whereas the East Coast weather is influenced by continental air from the landmass where it is colder, causing snow in winter. If the air would be moving straight down from high to low pressure areas without the influence of Coriolis acceleration, the weather patterns on both coasts of the continents would be similar.

For the Hadley cell too, the direction of motion of the wind at the surface is southward, Coriolis acceleration is to the east, and the reaction to it causes the winds to shift toward the west, creating the Northeast Trade winds. Trade winds are the steering winds of the tropical cyclones near the Equator; they determine the direction that the cyclones take as they travel westward.

Southern Hemisphere winds follow the same rules too, creating the Polar Easterly winds at the Polar cell, the Westerly winds within the Ferrel cell, and the Southeast Trade winds within the Hadley cell. See Figure 3.21 for the directions of these winds.

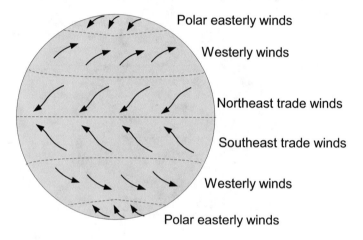

Figure 3.21: The prevailing winds shift due to Coriolis acceleration within each cell.

Note how the Easterlies and Westerlies create almost constant cycles in the Northern and Southern Hemispheres. These cycles have been used in sailing for millennia. They enable ships to sail from continent to continent. Similarly, depending on whether an airplane flies into a prevailing wind or against it, flight times can change significantly. Flying from San Francisco to New York usually takes less time than flying from New York to San Francisco, perhaps as much as one hour depending on different current conditions.

Figure 3.22 shows the general directions of the winds over the continents. Please remember that these are general directions. These wind directions are heavily influenced by seasons, bodies of water, mountains, and temperatures, therefore influencing regional and local weather patterns. Still, you can see how these winds influence the general weather patterns.

It is interesting that the same also happens on a smaller scale in your car. Next time you are driving in a car and the fan is on, try to notice what happens as

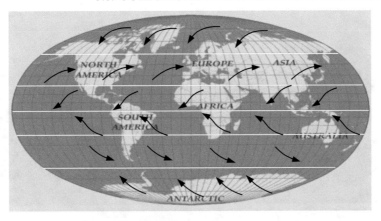

Figure 3.22: The general direction of the prevailing winds over the continents.

the car turns. If the blower is blowing the air to your face, as soon as you turn, the air's direction changes and you will not feel the air on your face. When you straighten out, the air blows in your face again. Why? Just like before, when the car turns, it becomes a rotating frame within which there is air moving. If you do the same cross product, with the vector representing the rotation of the car in the up-down direction and the blowing air in the direction toward you, the cross product of the resulting Coriolis acceleration will be to the left or right, deflecting the air sideways. When the car goes straight and the rotation vector is zero, the air moves directly toward you.

And now to our original question of why the weather of the West Coast is so different from the East Coast. As you may see in Figure 3.22, the general weather of California is heavily influenced by the Westerlies, which are commonly moist and warm due to the Pacific Ocean. Therefore, it does not snow until the mountains because the air is warmer and more humid, whereas, due to the same Westerlies that pass over a huge landmass, by the time the air reaches the East Coast, it is cold and therefore snows. The same is also true when the polar easterlies dip down and bring cold air from the north into those states. As you move down to the Southern Gulf states, both the Westerlies and the Easterly trade winds bring moisture to those states, so the air is warm and humid.

Of course, other local issues affect local winds and weather too. These include sea and land breezes during the day, mountains and valley breezes, and high mountain thermal flows that affect local winds (see Figure 3.23).

An important issue to notice is that the angular velocity (rate of rotation) of the Earth is very low (one revolution per 24 hours). Consequently, only long motions are affected by it over large

Figure 3.23: Rows of trees in San Luis Obispo, near the coast in California, all leaning to the east due to the influence of regular on-shore sea-breeze winds from the Pacific Ocean.

distances. Small motions we normally make like walking or throwing a ball are hardly affected by Coriolis at all. The idea, that the rotation of the water in a toilet bowl or when the water drains in a bathtub is due to Coriolis, is wrong; the water in a toilet rotates because it is deliberately designed with lateral openings to rotate the water as it is discharged to increase efficiency. The rotation of the water as it drains in a bathtub is due to the conservation of angular momentum, a subject that we have not covered here. But neither is due to Coriolis, and therefore, neither will necessarily go the opposite way in the Southern Hemisphere.

The importance of Coriolis acceleration varies based on latitude (the angle between the location and the Equator, defining north-south locations). Figure 3.24 shows the velocities of the surface winds within the Polar and Hadley cells again. Notice that since the winds follow the surface, the directions (line of action) of these velocities are not exactly the same, but follow the curvature of the Earth.

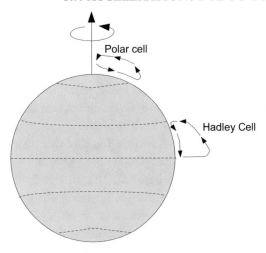

Figure 3.24: Surface wind directions of the Polar and Hadley cells.

Now let's look at the Coriolis acceleration of these two cells. Notice that as Equation (3.4) shows, Coriolis acceleration is the cross product of the angular velocity of the Earth and the wind velocity. But also as we saw earlier in Section 3.2.2, due to the nature of cross products, only the component of the velocity that is perpendicular to the angular velocity $\bar{\omega}$ counts; the cross product of the component of the velocity that is parallel to $\bar{\omega}$ is zero. As Equation (3.3) shows, the magnitude of the cross product is a function of $\sin(\theta)$, which is zero when two lines are parallel. As Figure 3.24 shows, the velocity of the surface winds in the Polar cell is nearly perpendicular to vector $\bar{\omega}$ representing the rotation of the Earth. However, the velocity of the surface winds in the Hadley cell is pretty close to parallel to $\bar{\omega}$. Therefore, the Coriolis acceleration and its effect is much more pronounced in the Polar cell compared to the Hadley cell near the Equator.

3.7 ACCELERATIONS DUE TO COMBINED MOTIONS

Gyroscopic motions are closely related to this subject and we will see a couple of applications where they are used later, but due to their complicated nature, we will not cover gyroscopic motions in this book. However, when there are multiple simultaneous rotations about different axes, they result in additional accelerations that cause interesting results. Since we have already discussed many of the elements related to these, let's go a bit further and also discuss these and their effect on the objects that many of us use regularly.

3.7.1 RIDING BICYCLES

One of the first things a child must learn for riding a bicycle is that turning the handle to the right or left will throw the rider to the side; instead he or she must *lean* to one side or the other to force the front tire to turn to one side. This is because, in addition to other factors such as the head-tube angle and the offset (called *rake* angle) and location of the center of gravity and friction, the combined rotations of the tire and the handle-bar create an additional acceleration similar to Coriolis acceleration that affects the turning (as mentioned earlier, this can be explained by gyroscopic motions, but we will not discuss that here).

While riding a bicycle there are two rotating frames, one turning within the other, causing acceleration. To see how this works, let's look at Figure 3.25. Imagine that a disk rotates about the x-axis as represented by vector ω_1. Now imagine that this disk is also rotating about the z-axis as represented by ω_2. As you can imagine, as time goes on, as a result of the rotation ω_2, vector ω_1 changes direction to ω_1' (and beyond). This change in direction, which in fact is the acceleration, is shown as $\Delta\omega$ (read delta-omega, meaning change in ω). Although we are not showing it here, this change is actually perpendicular to both of these vectors (along the y-axis) and its magnitude is $\omega_1\omega_2$. As we have seen before, this is the same as the cross-product of these two vectors. Therefore, the change $\Delta\omega = \bar{a}$ can be shown as:

$$\bar{a} = \bar{\omega}_2 \times \bar{\omega}_1. \tag{3.5}$$

This means that as one vector rotates due to a rotation, the resulting acceleration is perpendicular to both vectors. Now let's see how this translates into the bicycle ride.

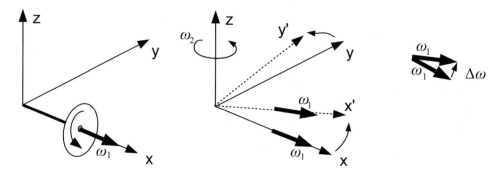

Figure 3.25: Rotations within a rotating frame cause acceleration.

Referring to Figure 3.26, Section 3.2.3, we see that the rotations of the tires of a bicycle can be represented by the vector ω_1 about the x-axis. If the handle-bar is rotated about the z-axis as ω_2 at the same time, there will be an acceleration perpendicular to both of these about the y-axis as $\bar{a} = \bar{\omega}_2 \times \bar{\omega}_1$. A reaction to this component of the acceleration due to inertia will tend to throw the rider to the right. Conversely, with the same ω_1 representing the rotation of the tires, if the rider leans to one side, there will be a rotation ω_2 along the y-axis, causing an acceleration along the z-axis that causes the handle-bar to rotate. As was mentioned earlier, there is a lot more to the total reaction of a bicycle, but this simple analysis shows how the bike reacts to rotations.

Figure 3.26: Rotations of a bicycle's tires and handle-bar and the resulting acceleration.

Why we can ride a bike without falling over: When the tires of a bicycle rotate, the rotation creates a vector perpendicular to the motion. This vector represents the angular momentum of the tire, and is a function of the weight of the tire, the way this weight is distributed (called moment of inertia), and how fast it rotates. Nonetheless, angular momentum is a vector whose direction follows the right-hand-rule. Although we have not discussed angular momentum yet, let it suffice to say that this vector likes to maintain its direction. In other words, changing the direction of this vector requires an attempt, an external torque. Otherwise, the direction and magnitude of this vector tend to remain the same. External factors such as friction eventually reduce the angular momentum. This is why a rotating wheel eventually comes to a stop. This is true for the direction too; the direction of an angular momentum vector tends to remain the same unless forced to change. This is called *preservation of angular momentum* and is an important subject in dynamics.

But what is important about this? The important thing is that due to this resistance to change, barring external influences, the direction of an angular momentum vector will not change. Therefore, as long as the tires of a bicycle are rotating (sufficiently) the direction of the vector resists changing. As a result, the tire will *not* fall over.

In some instances, if the rider brakes very hard, both tires may lock and stop rotating while they slide on the road before completely stopping. Unfortunately, this causes the bike to fall over because there is no longer any angular momentum. To prevent this, especially in motorbikes, the rear brakes are designed to never create forces large enough to lock the wheel. Since the contribution of braking force on the rear wheel is lower anyway, this generally does not significantly diminish the braking ability of the total system, but prevents the bike from falling over.

3.7.2   OSCILLATING FANS

An oscillating fan is in fact very similar to a bicycle. The rotating blades create a vector ω_1 perpendicular to the plane of the blades, as shown in Figure 3.27. Now imagine that you also turn on the oscillation mechanism that oscillates the fan to the right and left, as represented by a vector ω_2. Of course, the direction of ω_2 changes as the direction of oscillation changes. Therefore, the resulting acceleration $\bar{a} = \bar{\omega}_2 \times \bar{\omega}_1$ also changes direction. This means that as the fan oscillates either to the right or left, the forces generated by this acceleration will tend to push the fan's base forward, backward, or sideways. Note that here too, the magnitude of vector ω_1 representing the rotation of the blades is much larger than the magnitude of ω_2 representing the oscillations. Therefore, the resulting acceleration is relatively small.

Figure 3.27: The rotations and oscillations of a fan and their representations.

3.7.3   AIRPLANES

An airplane motions can be described by attaching a frame to it as shown in Figure 3.28. Although other conventions exist, the rotation of the airplane along an axis through its fuselage pointing forward is generally called *roll*. A rotation about an axis through the wings is called *pitch*, while a rotation to the left or right along a vertical axis pointing down is called *yaw*. As before, the three axes of the reference frame are mutually perpendicular. Conventionally, the x-axis is assigned to the roll axis, the y-axis is assigned to the pitch axis, and the z-axis is assigned to the yaw axis. Therefore, $\vec{x} \times \vec{y} = \vec{z}$. Notice how in this common convention for airplanes the positive direction of the yaw axis is downward.

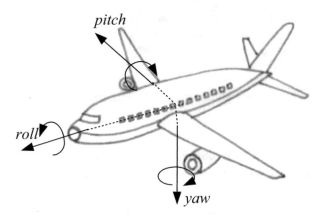

Figure 3.28: Motions of an airplane can be described through a frame attached to it.

First a word about how an airplane becomes airborne. An engineering principle called *Bernoulli's principle* indicates that when a fluid moves faster, its pressure drops. You can simplify this by looking at the total energy of a system, including both its potential and kinetic energies (see Chapters 1 and 2). Unless there is a net positive energy into the system, the total should remain the same due to *the conservation of energy* law. Therefore, as the kinetic energy of the system increases due to its higher speed, its potential energy (translated into its pressure) drops. This is used in many places and systems, for example in measuring the airspeed of an airplane or in the old carburetors that were used to mix gasoline with air and supply it to the engine. Take the airplane: a small pipe called a *pitot tube* is attached to the wing of the airplane and directly into the airstream. As the airplane flies, the air that is pushed into the pitot tube comes to a stop because the tube is closed at its opposite end. Therefore, the total kinetic energy of the air converts to potential energy, and as a result, pressure increases. As the plane goes faster, the pressure in the pitot tube increases. This pressure is measured and calibrated into the speed that the speed indicator shows. In the old-style engine carburetors a *venturi* was used to take advantage of the same principle. A venturi is essentially a tube whose diameter reduces at some point (see Figure 3.29). Since the same amount of gas or fluid passes through the smaller cross section, speed must increase, reducing pressure. Therefore, the pressure within the smaller cross section is lower than before or after. This was used to suck the gasoline from a small tank next to it, mixing it with the air and supplying the engine with the fuel-air mixture. Many other systems are also based on this principle. However, for an airplane, the increase in speed comes from the shape of the wing's cross section.

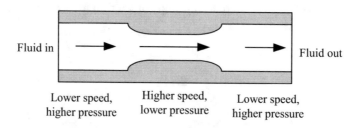

Lower speed,   Higher speed,   Lower speed,
higher pressure   lower pressure   higher pressure

Figure 3.29: As the fluid moves through a venturi, within the smaller cross section, its speed increases and its pressure decreases.

Looking at Figure 3.30 you will notice that as the airplane moves through the air, due to the asymmetric shape of the cross section of the wing which has a longer length on the top than it does on the bottom, the air has to travel faster above the wing than it does below in order to maintain continuity. As a result, the pressure above the wings is lower than the pressure below. This creates a positive upward force that floats the airplane. Notice that since the pressure above can never approach zero, the maximum difference between the pressure below and above is only a fraction of the total atmospheric pressure. However, since the wings are large, the pressure difference multiplied by the large area creates enough upward force to float the plane. By the way, you can test this by holding a piece of paper in your hands at one edge, letting the other edge hang freely. If you blow air over the paper it moves up. This is because the air moving over the paper has a velocity larger than the air below it. This reduces the pressure above the paper, lifting it.

We will discuss control surfaces and how airplanes' motions are controlled shortly, but first let's see how these surfaces work. A control surface is a portion of the wing or the rudder of an airplane which moves independently of it. Control forces are generated by moving the control surfaces into the air stream in different directions, resulting in controlled motions. Figure 3.31 shows a control surface that is lowered into the airstream. The air pushes against the surface and creates a force as shown, trying to "straighten" the obstruction. This force can be resolved into two forces, one horizontal, one vertical. The horizontal force is called *drag* and it works against the forward motion of the airplane, increasing demand from the engine to overcome it. The vertical force tries to move the surface and with it, the wing, upward. Similarly, if the control surface is

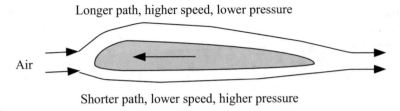

Figure 3.30: An airplane becomes airborne due to the pressure difference below and above the wings, caused by the increased velocity of air traveling above the wings. Due to Bernoulli's principle, as the air speed increases, its pressure decreases.

moved up into the airstream, the resulting force will be downward and the wing will move down with it too. These control surfaces control the motions about the three axes.

As you see, if the surface is lowered, the force will be upward. If the surface is lifted up, the force is downward. What happens if one surface is lowered while the opposite one is lifted up simultaneously? There will be a pair of forces, one downward, one upward. These two forces together create a torque (a *couple*) that causes rotation. This torque will rotate the airplane about the roll axis. In order to rotate the airplane along each axis a set of control surfaces are used, as shown in Figure 3.32, called ailerons, the rudder, and elevators.

Motion along the roll axis is controlled by control surfaces on the rear edge of wings close to the tips called *ailerons*. Ailerons move in opposite directions, developing forces that are also in opposite directions as the airplane moves through the air. The aileron that is up creates a downward force; the one that is down creates an upward force. As we saw in Section 3.2.2 these two forces will create a moment along the roll axis and will roll the airplane along that axis. Obviously, for controlled motions during flight, small forces are used to roll the plane slowly. But imagine a fighter airplane that rolls quickly to shun an incoming missile. The forces will be larger, requiring great skill on the pilot's part to control the plane.

Motion along the yaw axis is controlled by the rudder attached to the vertical tail fin, the large vertical control surface on the back of the airplane. The rudder moves to the left or right, creating a force to the right or left, rotating the airplane along the yaw axis. You may have noticed that the surface of a road is usually raised on one side along a turn. This is a lot more obvious in race tracks where race cars travel at very high speeds. This is necessary to keep the car on the road when during the turn, it is subject to centripetal acceleration (or what is referred to as centrifugal force which is really the reaction of inertia to centripetal acceleration as we discussed earlier). Similarly, airplanes are intentionally banked a little by rolling them along the roll axis during turns to prevent particles inside the airplane from flying outwardly (for example, if you have a glass of water on your tray in an airplane when it turns, without this banking, the glass will slide

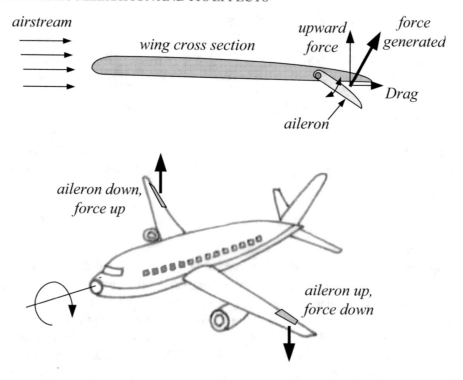

Figure 3.31: As the control surface is moved down into the airstream, a force is generated that can be resolved into drag and an upward force that lifts the wing on one side. The opposite happens if the surface is moved up into the airstream.

off the tray). In order to do this, the pilot usually combines the motions of the rudder with that of the ailerons to bank the airplane along the roll axis during a turn along the yaw.

Motion along the pitch axis is controlled by elevators. Elevators are really part of the rear tail surfaces (called *stabilizers*) that move upward or downward, either attached to the fuselage or attached to the top of the vertical fin. Alternately, a single control surface called *stabilator* may be used to do the same. Elevators and stabilizers are also used to level the airplane during the flight. In general, a pilot needs to make sure that the center of gravity of the plane is in front of the center of lift (usually indicated by a minimum and maximum distance), which is on the wings. In smaller airplanes before take-off the pilot makes sure that the luggage in the tail part of the airplane is moved around until the center of gravity is in front of the center of lift. This way, if the plane stalls, it will nose-dive (and not tail-dive). After the plane gains some speed, the pilot can try to level the airplane again and continue flying. Otherwise, as a plane stalls, it may crash.

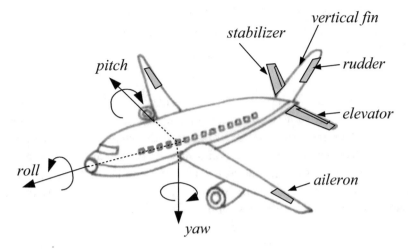

Figure 3.32: Control surfaces of an airplane. As a control surface is pushed into the airstream, the air presses against it, creating a force that attempts to push it back.

In some airplanes, the vertical fins and the stabilizers are combined into two tail surfaces that are at an angle relative to the wings (like a "v").

If you know how to make a paper airplane, make one. By bending both the wing tips up or down, one up and one down, or bending the rudder's tip, you can force the airplane to roll, pitch, or yaw in any combination.

How to Make a Paper Airplane: Figure 3.33 shows one way to make a simple paper airplane.

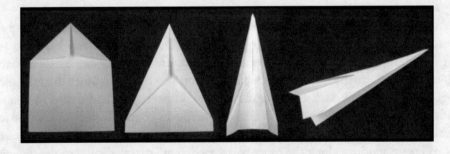

Figure 3.33: A simple way to make paper airplanes.

As we discussed earlier, any time there is motion within a rotating frame, there will be an additional acceleration component caused by the cross products of the two motions as shown by Equation (3.5) repeated here:

$$\bar{a} = \bar{\omega}_2 \times \bar{\omega}_1.$$

Like bicycles and fans, since the blades (or in the case of a jet engine, the turbines) rotate along the roll axis, there is a relatively large vector present in that direction. Whenever the airplane is rotated along the roll axis the same vector is present too. If at the same time the plane rotates along the pitch or yaw axes, there will be an acceleration component in the direction perpendicular to both. Therefore, anytime the pilot rotates the plane about the pitch axis, he or she has to also make a correction about the yaw axis, and whenever there is a rotation about the yaw axis, he or she needs to make a correction about the pitch axis. In most cases though, since the rotations are very slow, the resulting acceleration is very small and it may not be necessary to do much about it. However, in faster maneuvers or during take off (when the nose is pulled up by a rotation along the pitch axis) and landing, when these rotations are larger, corrections are necessary in some cases. Pilots are taught to look at the "ball in the race" instrument that indicates if the net acceleration vector is off to one side or not, and to "step on the ball" to keep this vector along the pilot's spine, thus keeping him or her straight in the seat, and not pushed to one side or the other. In larger airplanes the automatic control system of the airplane automatically takes care of these corrections.

You may have noticed that as airplanes touch the ground, the tires, which at that instant are stationary, rub against the tarmac until their speed is equal to the speed of the airplane, making noise and smoking. In this process, the tires wear as well. One common tendency is to suggest that the tires be rotated just before landing by attaching a small motor to each tire, therefore eliminating this rubbing and smoking. The problem is that since the tires rotate along the pitch-axis, as the pilot tries to correct the airplane's roll or yaw, additional motions are created along the other axes, making it difficult to control the airplane. There is a story that one airliner used tires that had small scoop-like extensions on the tires to force them to rotate when the landing gear was lowered before landing. Due to this phenomenon, the pilots used to cut the scoops out of the tires by a knife to prevent this rotation.

Auto-pilots are based on gyroscopes, which also involve the same ideas we have discussed. A gyroscope has a flywheel-type rotating mass, which due to its relatively large mass and high speed of rotation, creates a large angular momentum. This momentum resists a change in its direction. Therefore, anytime it is subjected to rotation in a direction other than the direction of the angular momentum vector, it resists the motion. Gyroscopes can be used in two ways, one to steady the motions of a system, another to automatically control its motions, called auto pilot. The same is true for ships and other water-vessels. For example, if a large gyro is mounted on a ship, it will resist motions along the other two axes that are perpendicular to the direction of the angular momentum vector. As a result, the gyro will steady the motion of the ship against waves. Additionally, based on Equation (3.5), since the gyro moves about an axis perpendicular to the motion induced in it, through sensors such as a potentiometer, a signal may be measured that can be used to move the control surfaces of the airplane to correct the induced motion. Therefore, except during take-off, landing, or emergencies such as sudden turbulences, while the airplane cruises at high altitudes the auto-pilot can be in control of the airplane.

3.7.4   ROBOTS

Robot manipulators, similar to the ones used in industry to manufacture and assemble parts and products, have multiple joints (degrees of freedom or DOF) that enable them to move to any position or orientation within their reach (see Figure 3.34). This is usually translated into a hip joint, a shoulder joint, an elbow joint, and three wrist joints (if the robot is a 6-DOF or 6-jointed robot. Fewer DOF or joints are also common, but they are not as versatile). Clearly, when all joints rotate together to move the robot to a new position or orientation, a similar situation like the bicycle, fan, or airplane exists; every rotation within another rotation creates an additional acceleration component perpendicular to the two rotations. Since the robot has up to six joints, it is possible that joints 1 and 2 may be moving simultaneously. Therefore, there will be an acceleration $\omega_1\omega_2$. Now suppose that joints 1, 2, and 3 move simultaneously. In this case, there will still be an acceleration component caused by the rotation of ω_2 (the shoulder) within ω_1 (the waist) as $\omega_1\omega_2$. However, there will also be another acceleration component caused by the rotation of joint 3 (the elbow) within the waist as $\omega_1\omega_3$, but since joint 3 also moves within joint 2, there will be an acceleration $\omega_2\omega_3$. Therefore, the total acceleration includes all three components. Similarly, if all six joints move together, since joints 2, 3, 4, 5, and 6 all move within joint 1, there will be acceleration components $\omega_1\omega_2$, $\omega_1\omega_3$, $\omega_1\omega_4$, $\omega_1\omega_5$, $\omega_1\omega_6$, and since joints 3, 4, 5, and 6 move within joint 2, there will be accelerations $\omega_2\omega_3$, $\omega_2\omega_4$, $\omega_2\omega_5$, $\omega_2\omega_6$, as well as $\omega_3\omega_4$, $\omega_3\omega_5$, $\omega_3\omega_6$, as well as $\omega_4\omega_5$, $\omega_4\omega_6$, and $\omega_5\omega_6$. Each acceleration, multiplied by its own corresponding mass or moment of inertia produces a Coriolis force that has to be dealt with. In most robots that move slowly, these accelerations are small and can be ignored. But for fast-moving robots, these can become significant and must be considered in the design of the robot.

To experience the same phenomenon yourself, first turn about your waist while your arm is stretched outward. Next move your arm up and down while still stretched. Then move your arm up and down while you rotate about your waist. You will notice how you feel an additional force against your arm, stemming from the Coriolis-type acceleration.

3.7.5   MOVEMENTS OF A SPACECRAFT IN SPACE

The following excerpt is from NASA's Gemini-VIII spacecraft mission journal, written by astronauts Neil Armstrong and David Scott (see https://www.hq.nasa.gov/alsj/alsj-Gemini VIII.html).

> After station-keeping for about 36 minutes, docking with the Gemini Agena Target Vehicle was accomplished. The final docking maneuver was begun when a distance of about 2 feet separated the two vehicles. A relative velocity of about three-fourths of a foot per second was achieved at the moment of contact. The nose of the space-

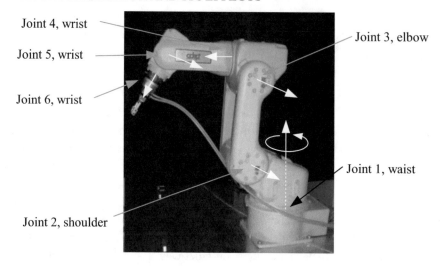

Figure 3.34: A robot and its joint movements.

craft moved into the docking adapter very smoothly and the docking and rigidizing sequence took place very quickly and with no difficulty. The docking sequence was completed at 6:33:22 ground elapsed time, with the two vehicles rigidized together.

For a period of 27 minutes after docking, the stability and control of the docked vehicles was excellent. At approximately 7:00:30 ground elapsed time, the crew noted that the spacecraft-Gemini Agena Target Vehicle combination was developing unexpected roll and yaw rates. The command pilot was able to reduce these rates to essentially zero; however, after he released the hand controller, the rates began to increase again and the crew found it difficult to effectively control the rates without excessive use of spacecraft Orbital Attitude and Maneuver System propellants. In an effort to isolate the problem and stop the excessive fuel consumption, the crew initiated the sequence to undock the spacecraft from the Gemini Agena Target Vehicle. After undocking, the spacecraft rates in roll and yaw began to increase, indicating a spacecraft problem which the crew attempted to isolate by initiating malfunction-analysis procedures. When the rates reached approximately 300 degrees per second, the crew completely deactivated the Orbital Attitude and Maneuver System and activated both rings of the Reentry Control System in the direct-direct mode. After ascertaining that spacecraft rates could be reduced using the Reentry Control System, one ring of the system was turned off to save fuel for reentry and the spacecraft rates were reduced to zero using the other ring. The crew continued the malfunction analysis and isolated the problem area to the No. 8 thruster (yaw left-roll left) in the Orbital Attitude and Maneuver System. The circuitry to this thruster had failed to an "on" condition.

This report relates to many of the principles that we have discussed in this chapter. To better understand these, let's start with thrusters and their role.

Large rockets are used to apply large forces (and acceleration) on spacecraft for rapid movements, but small motions and rotations are accomplished by firing small thrusters for short periods of time. Imagine a spacecraft, schematically shown in Figure 3.35, moving in space. Also imagine two pairs of opposing thrusters attached to it in one plane as shown. If thrusters A and B are fired simultaneously (usually for a very short time), the spacecraft will accelerate in the direction shown due to the force exerted by the thrusters. To slow down or return the spacecraft to its previous speed, thrusters C and D are fired simultaneously to exert similar but opposing forces to the craft.

To rotate the craft in the counter-clockwise direction, thrusters C and B are fired simultaneously. In this case, the forces of these thrusters are opposing and cancel each other, and consequently, the summation of forces is zero and the craft will not accelerate linearly. However, since these forces create a torque in the counter-clockwise direction, there will be a torque in that direction, resulting in a rotation. Similarly, to rotate the craft in a clockwise direction, thrusters A and D are fired simultaneously. In either case, to slow down the rotation or to stop it, the opposite pairs of thrusters are fired.

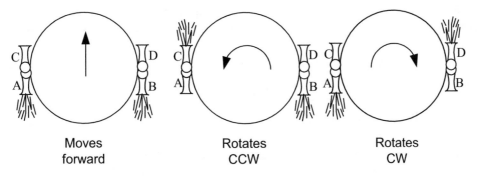

Figure 3.35: Depending on which pair of rocket thrusters is fired, the spacecraft may move in one direction, rotate clockwise, or rotate counterclockwise.

Since a spacecraft is free to move in three dimensions, it must have a similar arrangement of two pairs of thrusters in each plane to allow controlling its motions along the x-, y-, and z-axis as schematically shown in Figure 3.36. A similar arrangement is used for maneuvering spatial movements of astronauts in space walks.

What happened with the spacecraft in the earlier story was that one of the thrusters had malfunctioned in the "on" position, and was consequently applying a torque to the vehicle and accelerating it to about 300 degrees per second, a very large value that induces dizziness in astronauts and can eventually destroy their vehicle. The astronauts had separated the two crafts to find which one was at fault. After ascertaining that the docking vehicle was the reason, they fired another series of thrusters that are used for controlling reentry to counter the malfunctioning

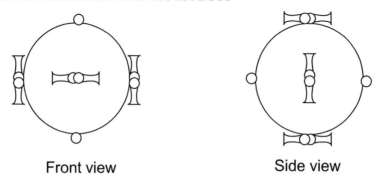

Figure 3.36: Pairs of thrusters are used in different planes (x-y, x-z, and y-z) to move a spacecraft along or rotate a spacecraft about its axes.

thruster. Eventually, the astronauts turned off the automatic system that was supposed to keep the vehicle running and took over the piloting of the vehicle themselves.

Notice that in spatial movements, since the object can rotate along all the axes, there are Coriolis accelerations along different axes as well. Therefore, quick rotations that increase Coriolis are more difficult to control.

Rocket propelled human flight systems follow the same rules, except that gravity is present. Therefore, thrusters or jet packs are only needed for lifting while gravity exerts a downward force. The rotations are accomplished by pairs of thrusters. But watch out for Coriolis acceleration.

Hopefully, you have noticed how all these issues are inter-related. Whether a bicycle, an airplane, the air coming out of a vent in your car, the weather, a fan, a robot, or a spacecraft, engineering principles govern how systems behave and react. Engineers use these principles in the design and analysis of systems that we use every day. Knowing them allows the engineer to not only create useful devices and systems, but also to protect users against adverse reactions that might occur as a result of these principles.

CHAPTER 4

Thermodynamic Cycles

Refrigeration, Air Conditioning, Engines,

and Power Cycles

4.1 INTRODUCTION

When I was a junior in an engineering college my uncle asked me, "Do you know how a refrigerator works? Can you repair one?" I replied yes, I know how it works. But whether I can repair one or not depends on a lot of other things. What I meant was that as engineering students, we learn thermodynamics, in which we study the principles that govern how a refrigeration cycle works, and based on that we can design the system. However, each company uses somewhat different sets of components to achieve about the same results. Based on experience with those components, you may or may not be able to fix a broken system or even recognize exactly what a part does; a certified technician can do that better than an engineer. But a technician cannot design the system or create a new one. The same is true with engines. You learn how an engine works and how to design it to ensure that it works properly, but as an engineer, you may or may not know how to fix it depending on your experience. To see this relationship and to understand why it is important to learn the basics and the principles of engineering let's look at refrigeration and power development systems and how the principles and the practical devices map into each other.

If you have access to a bicycle pump do the following exercise (if not, a simple balloon will do): Firmly place your finger at the output valve of the pump and press down on the plunger (down-stroke), pressurizing the air inside, and hold it (with the balloon, blow it up and hold the tip to prevent the air from escaping, but do not tie it with a knot). If you touch the body of the pump you will notice that it is a bit warmer (the balloon will most probably not get noticeably warm because of its size). Why do you think it is warm now?

This is because we perform "work" on the air to pressurize it (as was discussed in Chapter 3, work is force multiplied by distance. As a force moves, it does work, which also means that it adds energy to the system. In this example, we exert a force on the plunger to compress the air, and we move it in the same direction, doing work and adding energy to the system). The added work will increase the temperature of the pressurized air. The same happens when the air inside a tank is pressurized by a pump; the tank body's temperature rises a little because the air inside gets warmer. In mechanical engineering, the relationship between pressure, volume, and temperature of a gas can be expressed by an *equation of state*. The most common one for gases is called *ideal gas law*. Tables containing the detailed properties of real gases are also available. Equation (4.1) shows the ideal gas relationship between pressure, specific volume, and temperature. In this equation, temperature is the slave to the pressure and volume. R is called the *specific gas constant* and is known for different gases. For air, R is 0.2870 kJ/kg.K (kilo-Joules per kilogram Kelvin, where Kelvin is the absolute temperature, or K° = C° + 273. C is the temperature in degrees Celsius). In English units, R = 53.34 ft-lbf/lbmR (feet, pound-feet per pound-mass Rankin, where Rankin is the absolute temperature in English units or R° = F° + 460 where F is in Fahrenheit).

$$Pv = RT. \tag{4.1}$$

In this equation, P is the pressure, v is the specific volume (volume/mass), R is the gas constant, and T is the absolute temperature.

Due to the added work and as a result of Equation (4.1), the temperature of the air within the bicycle pump rises. We will see more about this shortly, but assuming that the ratio of the original volume of the pump to its final volume is 4:1 without any leaks, the final pressure can be about 7 times as much. Assuming a temperature of air before compression of 20°C (68°F), the final temperature can be as high as 237°C (460°F). You may ask why the pump warms up only a bit if the air itself is this hot? The answer is that the weight of the compressed air compared to the pump is very small. Therefore, although the temperature of the air increases a lot, the total energy is not much, increasing the pump's body temperature only a little.

A colleague of mine has designed a simple pump from glass in which the air can be quickly compressed about 20 times, raising its temperature to about 700°C (1290°F) and instantly combusting a small piece of paper that is placed at the bottom of the pump. Although the air inside becomes very hot, enough to burn the paper, the total energy is barely enough to warm up the body of the pump.

Now imagine that while you continue to hold the pressure, you let the bicycle pump (or the balloon) cool down. Here, while the air is under pressure, it cools down by losing its heat energy, and although the pressure drops somewhat, the air is at a higher pressure than when we started.

Next imagine that you release the plunger of the pump while you still hold the pump's outlet orifice (or let go of the balloon's tip to let the air out). Since the air is under pressure, it will push back the plunger, and as a result, the air's pressure returns (almost) to the original value before it was pressurized (in reality, the air is now doing some work and therefore loses some more energy).

Lastly, what happens at this point is that since the air returns to its original pressure and in the process has lost a net sum of energy, its temperature also reduces to something lower than what it was at the beginning; the pump will feel a little cooler at the bottom (and the balloon also feels cooler). Therefore, since it is now cooler than the environment, the air can absorb the heat from the environment and make it cooler. In this process, we forced the system to absorb heat from one environment, transfer the heat to another, and reject it there. The net effect is a lower temperature in one place and a higher temperature somewhere else.

This is exactly what happens in any refrigerator or air conditioning system. Neither of these systems "create" coldness; they just transfer the heat from one environment to another such that one becomes colder, one becomes hotter. Remembering our discussion about entropy in Chapter 1, is this not in the opposite direction of what natural systems would do to increase entropy?

Should we assume that the net effect is zero? In fact, we should not. Since we need to employ work (add energy) to compress the air and since all systems have friction (they waste energy, even if we were not compressing air but just moving the plunger), we add to the energy of the system which ultimately has to be rejected. Therefore, we will need to reject more heat at a higher temperature than we absorb at a lower temperature, making the higher-temperature source even hotter. In other words, as we discussed in Chapter 1 with entropy, the efficiency of the system can never be 100%. This is why if you run a refrigerator inside a room, even with its door open, the room will eventually become hotter not cooler; the total heat transferred to the room is equal to the heat from inside the refrigerator plus the electric energy used to run the compressor and the fans. What the refrigerator accomplishes is to keep its interior compartment cooler at the expense of a higher exterior temperature.

Now let's see what this means in engineering terms. First, notice that this is a cycle: we compress the gas that is at a particular temperature, let it cool down, then expand it which further cools it down, and in turn it absorbs the heat and warms up to its original state. We repeat the process. This is called a *thermodynamic cycle*. There are many different types of thermodynamic cycles, each with their own specific characteristics and applications, including cycles that translate into power development (such as in power plants), engines, and refrigeration systems. These cycles are usually described with graphs, including a graph of temperature versus entropy ($T - s$), pressure versus volume ($P - V$), and pressure versus enthalpy or energy ($P - h$). In this book we will use the pressure versus volume ($P - V$) diagram for its simplicity, even though it is somewhat limited in its usefulness.

Figure 4.1 shows the $P - V$ diagram for the first phase of this process. The x-axis shows the volume of the air at any state while the y-axis shows the corresponding pressure. Each of the isotherm lines show the relationship between the volume and pressure at a constant temperature. In this case, if the temperature of the gas is kept constant, as the pressure increases, the volume will decrease according the isotherm line. Now let's see how our process maps into this diagram.

Let's say we start at point 1 at a particular pressure and volume. In the case of the pump, this indicates the atmospheric pressure and the volume of the pump. Segment 1-2 shows the compression of the air; we compress the air in the pump, and as a result, its pressure increases, its

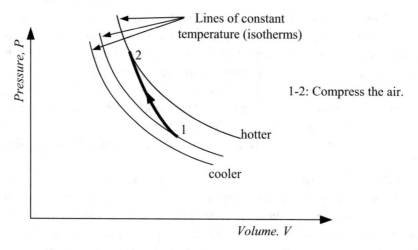

Figure 4.1: The pressure versus volume $(P - V)$ diagram of the first segment of the bicycle pump experiment.

volume decreases, and it becomes hotter. As shown, point 2 is at a higher pressure, lower volume, and at a higher temperature than point 1.

Segment 2-3 in Figure 4.2 shows the cooling of the air as we maintain the volume, but we let the air cool down (here, we are assuming that as the air cools down, its volume does not change. In reality, the volume decreases a little as it cools down). Notice how the line indicates the changes in the state of the gas. Its volume is (almost) constant, its pressure is lower, and the temperature is also lower as indicated by a lower-temperature isotherm, in this case the same as the original temperature.

Segment 3-4 in Figure 4.3 indicates the release of the plunger, where the pressure returns (almost) to the previous level and the volume is a little lower too (due to the decrease in temperature). However, notice that point 4 is at a lower temperature than point 3 (lines 1-2 and 3-4 follow a constant entropy line).

Segment 4-1 in Figure 4.4 is the absorption of outside heat energy into the system, which returns the system back to its original state. These four segments constitute a cycle that can be repeated. In the process, we transfer heat from one environment into another by applying external energy to the system, with a negative net effect.

4.2   REFRIGERATION CYCLE

The way a refrigerator works is very similar to the bicycle pump example, except that as an engineered device, it is designed to be much more efficient and to work continuously. Each part of

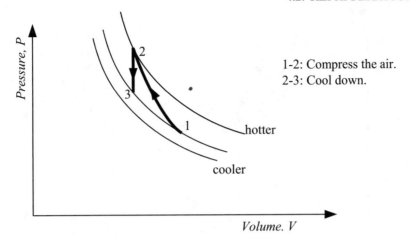

Figure 4.2: The pressure versus volume ($P - V$) diagram with the second segment of the bicycle pump experiment.

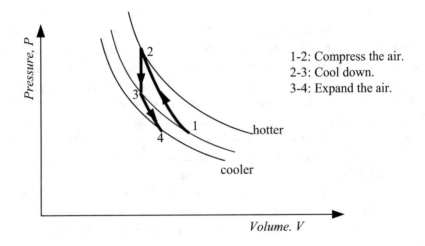

Figure 4.3: The pressure versus volume ($P - V$) diagram with the third segment of the bicycle pump experiment.

the cycle is accomplished by a particular component. Let's see what these components are and how they work and how the cycle differs from Figure 4.4.

Unlike the bicycle pump, where the medium was air, the medium in most refrigeration systems is a chemical with favorable characteristics such as boiling point and heat capacity. For decades, the refrigerant of choice was Freon-12, a chlorofluorocarbon (CFC), and it worked very

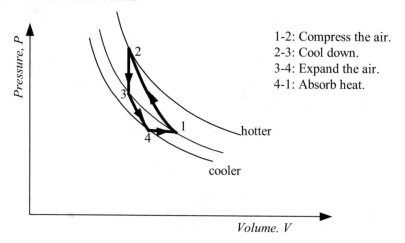

Figure 4.4: The complete pressure versus volume $(P - V)$ diagram of the bicycle pump experiment cycle.

well due to its physical characteristics. Freon-12 was also freely used as a propellant in sprays. However, since Freon had an adverse effect on the upper-atmosphere Ozone layer, it was banned in the mid 1990s. It was replaced with tetrafluoroethane (R-134a). Although R-134a works well too, due to the size of its molecules, it leaks more easily and needs to be replaced more often. It turns out that R-134a also has adverse effects on global warming (as much as 2,000 times more than CO_2) and is being phased out. A big difference between the bicycle pump example and a refrigeration system is that instead of air, which is always a gas, Freon switches states between gas and liquid. This helps in maintaining a desired and almost constant temperature in the refrigerator and freezer and increases the efficiency of the system.

Figure 4.5 shows a typical refrigeration system. The system consists of four operations; compression, condensation, expansion, and evaporation. These operations are accomplished by components (conveniently) called a compressor, a condenser, an expansion valve, and an evaporator.

The compression segment of the cycle is accomplished by a compressor. A compressor is a combination of an electric motor and a pump, both integrated together and hermetically sealed to prevent leakage. The function of the compressor is to compress the refrigerant. As a result, the refrigerant becomes a pressurized hot gas. In many cases, a small fan is installed next to the compressor to blow air on it to keep it cool; otherwise the heat may damage the compressor. Remember that in order to compress a gas, we need to do work on it, adding to the energy of the system. This work, supplied by the compressor motor, eventually turns into heat and is ultimately wasted. Figure 4.6 shows a typical compressor and cooling fan next to it. Typically, the compressor is in the rear-bottom part of the refrigerator and can be accessed from the back.

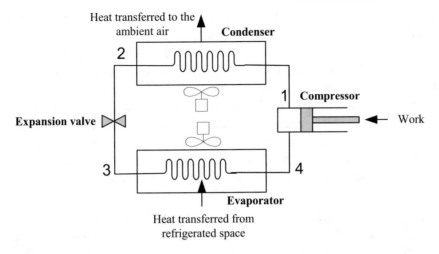

Figure 4.5: A typical refrigeration system and its components.

Figure 4.6: A typical compressor and cooling fan next to it.

The second component of the system is a condenser. In reality the condenser is a simple heat exchanger; it is a series of tubes that transfer the heat of the high-pressure hot refrigerant out to the ambient air. As a result, the compressed and overheated refrigerant cools down, and eventually becomes a liquid (and this is why it is called a condenser, because it condenses the superheated gas into liquid). Therefore, at point 2 in Figure 4.5, the refrigerant is in liquid form. To increase heat transfer from the condenser, it is possible to add a fan to blow air over the condenser and cool it down. In older refrigerators, the condenser was usually placed vertically on the back of the unit in order to take advantage of convection heat exchange; the warm air would simply rise and escape from behind the unit. In modern units, the condenser is usually under the refrigerator. In reality, the small fan that is used to cool down the compressor is designed to suck in the air from the front of the refrigerator at the bottom, pass it over the condenser and the compressor, and blow it out the back, in effect cooling both of them together. Condensers are very prone to dust accumulation that greatly reduces their effectiveness. Therefore, it is advisable to clean the condenser once in a while. It should be mentioned that for larger systems, water cooling, larger fans, and other assistive devices are added to remove larger heat loads. Figure 4.7 shows typical condensers in a household refrigerator (a) and in an industrial unit (b).

(a)   (b)

Figure 4.7: Typical condensers in refrigerators (a) and industrial units (b).

The third component of the system is an expansion valve. Typically, the expansion valve is a long capillary (very narrow) tube. When the liquefied refrigerant passes through it, due to the large pressure drop in the capillary tube, the high-pressure liquid loses its pressure and becomes a mixture of gas and liquid at low pressure. Just like the bicycle pump example, when the cooled refrigerant is allowed to lose its pressure, its temperature drops significantly. Therefore, at point 3 in Figure 4.5, the liquid entering the evaporator is very cold, and consequently, the evaporator will also be very cold. Figure 4.8a shows a typical expansion valve. In larger systems the expansion valve can be an actual valve, which similarly reduces the pressure of the liquid as it passes through (Figure 4.8b).

The fourth component of the system is an evaporator. In reality, the evaporator is also a simple heat exchanger just like the condenser, and is similarly made of tubes. The cold refrigerant mixture of gas and liquid absorbs the heat of the refrigerator and boils into gas, which is then sent back to the compressor. In this process, the heat of the refrigerated area is absorbed and

(a)   (b)

Figure 4.8: Typical expansion valves.

transferred to the outside. To increase the effectiveness of the refrigerator in modern systems a fan blows air over the evaporator and then into the freezer which is typically at about $4°F(-15°C)$. The refrigerator area is cooled through the freezer air and is typically at $35°–40°F (1–4°C)$. In most systems, the evaporator is behind the freezer area and cannot be seen. In older refrigerators, the evaporator coils were embedded into the freezer box area.

Figure 4.9 shows a typical thermodynamic $P - V$ (pressure vs. volume) refrigeration cycle. As expected, it includes the same compression, condensation, expansion, and evaporation segments. Segment 1-2 in Figure 4.9 shows the thermodynamic representation of compression. During compression, pressure increases, volume decreases, and temperature increases. Although not entirely accurate, it is usually assumed that the entropy of the system remains the same during this operation. This is not an important issue for our discussion here, as we have not really studied this subject. However, the assumption helps us determine how this segment behaves in the $P - V$ diagram. All points under the curve are mixtures of gas and liquid. All points to the right of the dome are gas. Therefore, at point 2, the refrigerant is in gas form. Segment 2-3 shows the condensation, where the refrigerant condenses by losing heat and becomes a mixture. In this process, pressure remains the same, but since the gas liquefies, the volume decreases. Segment 3-4 is the expansion, where volume increases a small amount, but pressure drops and temperature decreases. Segment 4-1 is evaporation, where the liquid evaporates by absorbing the outside heat at constant pressure and its volume increases.

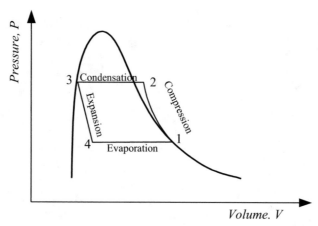

Figure 4.9: Thermodynamic representation of the refrigeration cycle.

While designing a system, which is what many engineers do, the designer has to choose components of the system so that they collectively work as desired. The thermodynamic cycles such as Figure 4.9 allow the engineer to design the system, pick appropriate specifications (such as temperatures, pressures, volumes, etc.) and choose appropriate components that work together and produce the desired results. This includes the size, capacity, and power of the compressor, the size of the fan, the dimensions of the condenser and evaporator, the ranges of temperature and pressures, and so on. Without these thermodynamic tables and material behavior charts, it would be impossible to design a system that is efficient and works well.

It should be mentioned here that modern refrigerators have other added features. For example, huge layers of ice would form on the freezer walls of old refrigerators that doubled as an evaporator. This is because when the air cools and its temperature drops the moisture in the air condenses to water and freezes over the cool surface. To prevent this, about every 22 hours, refrigerators switch off and the freezer walls are heated slightly for a short time to thaw the ice, which drops down into a tray at the bottom of the refrigerator. The water eventually evaporates into the outside room. This action keeps the freezer ice-free, but uses energy to heat the freezer wall and melt the ice, but also to cool it down again.

Air-conditioning systems are essentially the same as refrigeration systems, except that the components may be put together somewhat differently to cool down an environment instead of the limited volume of the refrigerator. These systems also include a compressor, condenser, expansion valve, and evaporator. The heat of the system is transferred to the outside by a fan, whereas the air from the room is sucked in by a fan, blown over the evaporator to cool down, and returned to the same environment. In this process, since the air is cooled down, some of the moisture in the air condenses on the evaporator, and therefore the humidity of the air is reduced too. In the summertime, this is a good thing because moist air feels warmer. Consequently, the air feels better because it is cooler and also drier. However, like refrigerators, the condensed water has to be drained. You may have noticed that in many air-conditioning systems, it appears that the unit is leaking. That is in fact condensed water and not a leak. The same is true in automobile air-conditioners. Other than these differences, an air-conditioning system and a refrigeration system are thermodynamically very similar.

4.3   SPARK-IGNITION POWER CYCLE

A spark-ignition cycle approximates the cycle of power development by an internal combustion engine with spark plugs. This is also similar to what is referred to in thermodynamics as an *Otto Cycle* which is an ideal cycle (an ideal cycle is approximate. Real cycles differ somewhat from ideal cycles. But to learn the principles, we always start with an ideal cycle, then modify the cycle to a more realistic model). Conversely, a compression ignition cycle approximates a diesel engine, where the air is compressed much more and consequently, it becomes much hotter to the point that when the fuel is injected into it, it explodes and burns without the need for a spark plug. We will discuss the differences between these two engines later.

An internal combustion engine in general refers to any type of engine in which the combustion of the fuel and air within a closed environment produces the gases that generate the mechanical work, and includes regular gasoline engines, diesel engines, rotary engines, and jet engines. Conversely, steam engines are not internally combusting engines; in steam engines, the fire is outside of the engine and instead, combustion products boil water into steam in a boiler and the steam is used to power the engine. Common gasoline and diesel engines are called *reciprocating IC engines* because the piston reciprocates (moves up and down) in a cylinder, rotating a crankshaft that is connected to it via a crank and a connecting rod. Wankel (rotary) engines and jet engines do not reciprocate (in fact, they do not have pistons and connecting rods and cranks); instead Wankel engines have rotary 3-sided rotors that revolve within a chamber, and jet engines have compressors, combustion chambers, and turbine rotors that always rotate. We should remember that in this section our discussion revolves around reciprocating internal combustion engines even if we just refer to them as IC engines.

First let's see how an engine works, then we will look at the thermodynamic cycle representing it. It should be mentioned here that there are two types of gasoline reciprocating internal combustion (IC) engines—2-stroke and 4-stroke. As we will see shortly, 2-stroke engines are less efficient and more polluting than 4-stroke engines. Four-stroke engines are cleaner, more efficient, and vastly more popular, but as we will see, they develop power in every other cycle. Except in specific cases such as small motorbikes, model airplane engines, some lawn-mower engines and the like, almost all engines are 4-stroke.

4.3.1   4-STROKE ENGINES

Figure 4.10a shows the schematic of a one-cylinder, 4-stroke, spark-ignition reciprocating internal combustion engine (notice all the qualifiers that are used to define it). In kinematics, this is called a slider-crank mechanism (Figure 4.10b) because it consists of a slider (the piston) and a crank, connected together by a coupler. However, unlike the slider-crank mechanism where the slider simply slides on a surface, in the internal combustion engine the piston oscillates inside a cylinder with a cylinder head which enables it to compress the air inside. The engine schematic shows the engine block and cylinder head, the piston, the connecting rod (coupler), the crank and crankshaft, the spark plug, the fuel injector, and two valves on the top. One valve is on the intake manifold and allows the outside air to be sucked into the engine; the other is on the exhaust manifold and allows the burned gases to be pushed out to the atmosphere.

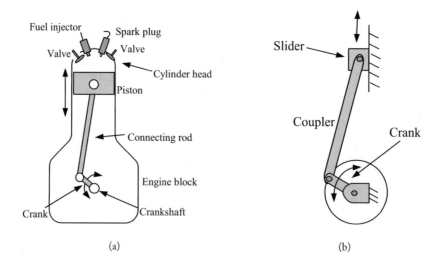

Figure 4.10: A single-cylinder reciprocating internal combustion engine and a slider-crank mechanism.

In 4-stroke engines the complete cycle occurs within two complete rotations of the crankshaft (720°), causing the piston to move up and down twice within the complete cycle (Figure 4.11). These are called the *intake*, *compression*, *power*, and *exhaust* strokes. Let's assume that the piston is at the top (called *top-dead-center* or TDC) and is at the beginning of a cycle (intake stroke). At this point the intake valve is open, and as the piston moves down, it sucks in filtered air, as shown in Figure 4.11a. In older cars, the air would move through a device called a carburetor. Carburetors are no longer used in cars, but they are still around in older cars (even into the 90s). As discussed in Chapter 3, when fluids or gases move faster, their pressure drops. A carburetor has a Venturi that, due to its reduced cross section, increases the speed of air, dropping the pressure. The pressure difference through the Venturi sucks in some fuel from a small tank at the side of the carburetor, causing fuel (gasoline) to be mixed into the air-stream. In fact, for this reason, the faster the engine rotates, the better the fuel mixes with air. The fuel-air mixture continues to the cylinder. In newer engines, an injector is used to inject a precise amount of gasoline into the cylinder as the piston moves down, mixing with the air (other possibilities exist). When the piston is almost at the bottom (called *bottom-dead-center* or BDC), the intake valve is closed.

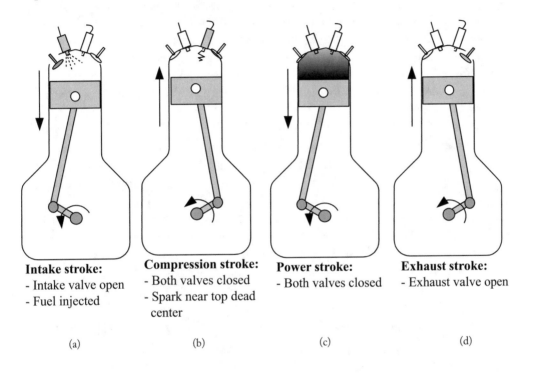

Intake stroke:
- Intake valve open
- Fuel injected

(a)

Compression stroke:
- Both valves closed
- Spark near top dead center

(b)

Power stroke:
- Both valves closed

(c)

Exhaust stroke:
- Exhaust valve open

(d)

Figure 4.11: The four strokes of a 4-stroke gasoline engine.

The second stroke is compression. As the piston moves up, since both valves are closed and the fuel-air mixture is trapped inside the cylinder, it compresses causing both the pressure and temperature to rise (Figure 4.11b). In most gasoline engines, the compression ratio, the ratio of the air volume at the beginning and end of this stroke (V_1/V_2), varies between about 8 and 11. A compression ratio of about 11 raises the temperature of the air to the point that it will require premium gasoline with higher octane; otherwise, the fuel-air mixture may ignite at an improper time, potentially damaging the engine (called *detonation* or *knocking*). At compression ratios of about 8 the temperature is still low enough to not combust prematurely; therefore regular gasoline can be used.

It should be mentioned here that the ignition, combustion, and burning of the fuel-air mixture is a very complicated and involved issue that is beyond the scope of this book. We simplify this process greatly here. In fact, one of the major issues in an internal combustion engine class in engineering programs is the process of combustion and its mechanisms.

Gasoline Octane Number: The gasoline octane number relates to the temperature at which gasoline auto-ignites in an engine. The higher the octane number, the higher this temperature will be. Gasoline octane numbers vary between 87 and 91. Gasoline with octane number 87 is called regular, 89 is medium grade, and 91 is premium gasoline, not in quality but in the auto-ignition temperature. The auto-ignition temperature of gasoline is between 246–280°C (475–536°F) depending on its grade when measured in open air. In engines, the temperature at which the gas may ignite prematurely is much higher, at about 747 ± 22°C (1380 ± 40°F) [1]. These numbers do not relate to the quality or cleanliness of the gasoline at all. Higher octane in a fuel is achieved by mixing different hydrocarbons and by adding chemicals to the gasoline that increase its ignition temperature, and therefore, make it more expensive. However, all grades have the same heat energy value. If your engine's compression ratio is higher, say 11, it will require higher octane gasoline. If it is about 8-9, it can safely work with regular 87-octane gasoline.

Should you use the higher octane gasoline in an engine with compression ratio of about 8-9 as some suggest, thinking it is a better gasoline? The answer is no. Higher octane gasoline will not provide more energy, will not burn better or cleaner, and will not keep your engine cleaner. Therefore, for more money, you will get the same result. Using higher grade gasoline in a car that does not require it will make no difference except in unnecessary higher cost.

The only exception is when an engine has carbon deposits in it that cause it to continue to turn even when turned off, or if the carbon deposit causes pre-ignition. This happens to older engines in which after a long period of time, carbon deposits accumulate in the cylinder; when the engine is hot, the deposits cause the air-fuel mixture to combust before the piston gets to the ignition point when the spark plug sparks. This premature ignition causes the mixture to burn prematurely, increasing pressure to unsafe levels. This may damage the engine permanently. Additionally, in some engines the same carbon deposits cause the mixture to ignite although the engine is turned off, continuing to rotate. Higher octane gasoline may improve the situation by decreasing the probability of the mixture igniting prematurely. Otherwise, continue to use the grade that the manufacturer recommends.

Most typical engine management systems found in modern cars have a *knock sensor* that monitors if the fuel is pre-igniting. In modern computer controlled engines, the ignition timing will be automatically altered by the engine management system to reduce the knock to an acceptable level. However, this should not be a reason to use regular gasoline in a high-compression engine that requires premium grade either.

The third stroke is the power stroke. Within the two rotations of the crankshaft, this portion is the only one that actually delivers power to the engine. As the piston nears the top-dead-center, the spark plug is fired to generate a strong spark within the fuel-air mixture, causing it to combust and burn quickly as the piston clears the top-dead-center. Combustion creates a very high-pressure mixture, which when multiplied by the area of the piston, translates to a very large force, pushing down the piston in its down-stroke and generating a large torque at the crankshaft. Both valves remain closed during this stroke.

Just a note here. Do you remember the definition of work (force multiplied by distance)? Here, the force of combustion gases on the piston pushes it down, therefore moving the piston. Consequently, we have a force that displaces, creating work, or energy. The force is not constant, therefore the rate of work generated is also not constant.

Finally, the fourth stroke is the exhaust. As the piston starts to move up again, the exhaust valve is opened, allowing the piston to push out the hot, burned gases, almost completely clearing the cylinder of the spent fuel-air mixture and preparing it for the repetition of the first stroke, sucking in fresh air as the exhaust valve is closed and the intake valve is opened. The cycle repeats until the engine is shut off.

What is the purpose of higher compression ratios if they require more expensive gasoline? In general, the higher the compression ratio, the better the efficiency of the engine. Engine efficiency percentages range from the 20s to the low-30s. As the compression ratio is increased, it compresses the same air more compactly, increasing its temperature and reducing its volume. As a result:

1. The mixture burns better at higher temperatures.

2. Because compression and combustion happen in a smaller volume, the fuel mixes better and burns more quickly and completely.

3. Higher pressures produce larger forces.

4. More power is squeezed from the combustion gases.

When an engine operates at high altitudes, since the air is thinner, less air enters the cylinders. As a result, compression pressure is lower and the engine delivers less power. Turbochargers are used to increase the intake pressure and push more air into the cylinder, especially at higher altitudes.

For ideal gases, the ratio between the pressure at bottom-dead-center P_1 and top-dead-center P_2 can be calculated as a function of compression ratio r_c (ratio of volumes at bottom-dead-center V_1 and top-dead-center V_2 or $r_c = V_1/V_2$) as:

$$\frac{P_2}{P_1} = \left(\frac{V_1}{V_2}\right)^n, \tag{4.2}$$

where $n \approx 1.4$ for ideal gases. Therefore, we can see that for a compression ratio of 8, the pressure ratio will be about 18 whereas for compression ratio of 11, the pressure ratio increases to 28, a significant increase.

Except in specific applications such as 1-cylinder lawn mower or power tool engines and some small cars with 2- or 3-cylinder engines, most automobile engines have at least 4 cylinders. Five, 6, 8, and even 12 cylinders are also common. In multiple-cylinder engines each cylinder and piston combination operates exactly as mentioned before. However, all connecting rods are attached to a common crankshaft. Consequently, the movements of the pistons are all coordinated. For example, in a 4-cylinder engine, if one piston is at top-dead-center and is at the beginning of its intake stroke, another piston might be at bottom-dead-center and at the beginning of its upward motion to compress the fuel-air mixture. A third cylinder may be at TDC and at the beginning of its power stroke, while the fourth is also at BDC and ready for its exhaust stroke. The same sequence continues between all four, and as a result, in every stroke, one of the cylinders is at its power stroke.

As mentioned earlier, in 4-stroke engines, the total cycle for each piston requires two rotations of the crankshaft or 720°, and therefore, each stroke is one-quarter of this, or 180°. Consequently, in a 4-cylinder engine, one power stroke occurs at every 180° of the crankshaft rotation. As a result, the multiple-cylinder arrangement will make the output power much smoother than if there were only one larger cylinder. For a 6-cylinder engine, the power stroke is at every $720°/6 = 120°$, and for an 8-cylinder engine, it is at every $720°/8 = 90°$. Therefore, even for the same size engine, a 6 or 8-cylinder engine will run much more smoothly than 4-cylinder engines. Notice that since one cylinder at a time produces power the output is more uniform, whereas if they were all arranged to fire simultaneously (which is possible if they are all connected to one crank), the output would vary much more, causing a much rougher ride.

It should be mentioned here that it is extremely crucial that the valves close and open at exact proper times. The valves are opened and closed by pear-shaped cams on a camshaft (Figure 4.12). To coordinate the valve timing with the position of the pistons, the camshaft is run directly by the crankshaft through gears, a timing belt, or a timing chain at half the speed. As the camshaft rotates, the cams on it turn and open the valves; springs close the valve. All these motions require work, which comes from the crankshaft (part of the power developed by the power stroke). Therefore, part of the energy of the engine goes into running its internal parts. Later-model engines may have two valves for intake and two valves for exhaust in order to speed up the process of intake and exhaust. Consequently, a 4-cylinder engine may have 16 valves (specified as DOHC engine for double-overhead-cam engine). In these engines, since the four valves can be in four corners of the combustion chamber in the shape of a plus sign, there is room in the middle for the spark plug. As a result, the combustion starts in the middle of the chamber with equal distance to the perimeter, more completely burning the fuel-air mixture. Therefore, these engines produce less pollution and are more efficient too. Figure 4.13 shows the same cylinder-head block with the valve arrangement and the camshafts on it.

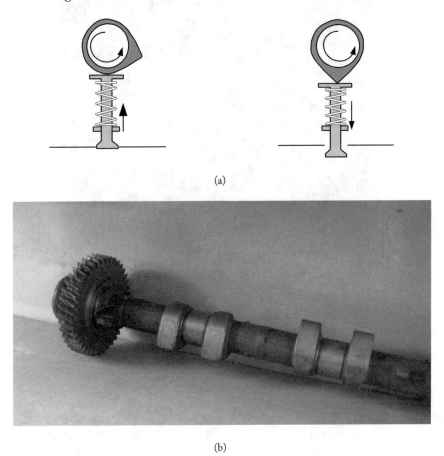

(a)

(b)

Figure 4.12: A cam opens and closes the valve as it rotates.

Exhaust valves

Spark plug opening

Intake valves

Figure 4.13: The valve arrangement and the camshaft on a cylinder-head block.

To facilitate quicker passage of air in or out of the cylinder, the designer of the engine should want to open the valves as much as possible. However, there is a limit to how much this can be because when the piston gets to top-dead-center, the remaining volume is very small and we do not want the piston to run into the valves. However, there is an additional concern. Like any other mechanical device, it is possible that the timing belt or chain may break. In that case, some valves may remain open while the piston continues to travel to the top, eventually running into them. This can be disastrous to both the piston and the valves, and should be avoided at all costs. This is why manufacturers recommend that every so often, the timing belt be replaced *before* it breaks. Others use a timing chain, which in general can last much longer without failure.

Additionally, it is possible to design the engine in such a way to ensure that the pistons and the valves will *not* collide at all. These engines are referred to as *non-interference* engines versus interference engines in which the pistons and valves may collide if the timing belt breaks. In non-interference engines, although the engine stops working if the belt breaks, it remains safe and as soon the belt is replaced, the engine can be used again—a minor repair that can be done easily. In interference engines, if the timing belt breaks, the engine may require major overhaul. Find out what the engine in your car is and how often you need to replace the timing belt.

As you may imagine, there is a very large amount of heat generated in an engine, a large portion of which must be transferred to the environment; otherwise, the engine parts will overheat and will be damaged. In order to keep the engine cool, most engines have a water-cooling system and a radiator that transfers excess heat to the environment. To do this, there are water passages throughout the engine block. A water pump forces the water around the cylinders and the engine body, and later, through the radiator. With the aid of a fan, the radiator transfers the heat out. To keep a constant range of temperatures, a thermostat is used to stop the flow when the coolant is cold, and open when it gets hot.

Alternately, some engines, including some automobile engines as well as airplane engines, are air-cooled. In this case, the flow of air over the engine fins cools the engine.

The other major issue in engines is friction between the contact surfaces, including the piston and the cylinders, the connecting rod and the cranks, and the cranks and the crankshaft. The friction causes additional heat that must be removed as well. To reduce friction and to cool down these engine parts they are constantly lubricated with engine oil. The crankcase, the big reservoir at the bottom of the engine block, is filled with oil. An oil pump, sometimes inside the crankcase, pumps the oil between these contact surfaces and also splashes some oil onto the inner surfaces of the cylinder when the piston is at the top, lubricating the contact surface as the piston slides down. Of course, the pumping of the oil also takes away a little more of the engine power.

4.3.2   2-STROKE ENGINES

So far we have studied 4-stroke engines. However, as mentioned earlier, there are also 2-stroke engines that are used with simpler systems such as motor bikes or model airplanes. In this case, all necessary parts of the cycle have to happen within two strokes (one complete revolution of the crankshaft) or within 360°.

There are advantages and disadvantages to 2-stroke engines. One advantage is that since the power stroke happens at every 360°, we should expect that the power development per cycle is more dense in a 2-stroke engine than in a 4-stroke engine, where the power development is once every 720°. But due to other inefficiencies of the 2-stroke engine, this ratio is *not* twice as much. Another advantage of a 2-stroke engine is that since it lacks the similar valve arrangement necessary in a 4-stroke engine, it is usually much simpler with fewer parts. Therefore, it is an appropriate design for model airplanes and other applications where cost, space, and weight are important issues. However, as we will see, due to their construction, these engines are more polluting and wasteful, and due to the lack of an oil pump, they require that the oil be added to the fuel. Therefore, the engine burns a mixture of oil and gasoline, which makes it even more polluting. It is expected that the mixture lubricates the engine as it goes through the system.

Two-stroke engines do not have intake and exhaust valves. Instead, there are two openings on the lower part of the cylinder body that are normally closed when the piston is up and covers them, and open as the piston moves down. In 2-stroke engines, the crankcase is also closed except through a valve, as depicted in Figure 4.14. When the pressure in the crankcase is lower than the outside, it opens and air is entered into the crankcase; when the pressure in the crankcase is higher, it simply closes. Therefore, as the piston moves up and creates relatively lower pressure in the crankcase, air is sucked in. As the piston moves down, the valve closes and the air that is trapped in the crankcase is pushed up into the cylinder through the intake opening, all the while sucking a little gasoline-oil mixture with it into the cylinder (see the previous discussion about venture effect). Unlike 4-stroke cycles, more than one thing happens simultaneously during each stroke of the 2-stroke cycles. Therefore, we need to start at some arbitrary point, follow all that happens, and eventually end at the same point in order to see how this engine works.

Imagine that the piston is moving down and is close to its bottom-dead-center as in Figure 4.15a. By this time, the intake valve in the crankcase is closed due to the increased pressure in the crankcase, the exhaust port is opened and the consumed fuel-air mixture is mostly out, and as the piston continues its downward motion, the intake opening on the cylinder opens as well, and the somewhat-compressed fuel-oil-air mixture in the crankcase is pushed into the cylinder.

As the piston starts its upward motion (Figure 4.15b), it closes the intake and exhaust openings, opens the crankcase intake valve bringing new fuel-oil-air mixture into the crankcase for the next cycle, and compresses the mixture in the cylinder until it reaches near the top-dead-center. At that point, a spark combusts the mixture, starting the downward power cycle.

Once again, as the piston moves down, it closes the crankcase intake valve, opens the exhaust and the intake openings in short sequence, and repeating the cycle until the engine stops. Clearly, before all the exhaust gases escape, a new fuel-oil-air mixture starts entering the chamber and mixing with it. This reduces the efficiency of the engine and also lets some of the new unburned mixture out the exhaust, polluting the air. As was mentioned earlier, these engines are simple, with fewer moving parts, and with more frequent power cycles, but are more polluting and less efficient.

4.4 THERMODYNAMIC REPRESENTATION OF THE SPARK-IGNITION POWER CYCLE

Similar to the refrigeration cycle, where we compared the actual work of the system with its thermodynamic representation, we can do the same for power cycles. This can help engineers

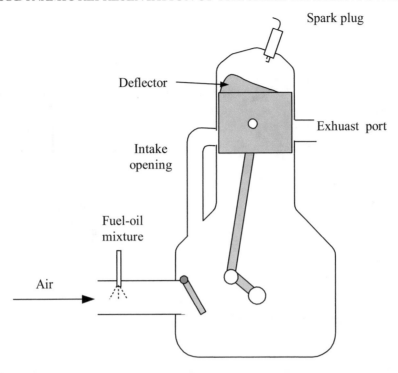

Figure 4.14: Schematic of a 2-stroke engine.

design engines with desired specifications, calculate the power developed by different size engines, and help them design more efficient and better engines. It is crucial for an engineer to work with thermodynamic representations.

Once again, it is very useful to also study the temperature-entropy $(T - s)$ diagram as well as the pressure-volume $(P - V)$ diagram. However, since we have not studied entropy as a tool, we will skip the $(T - s)$ representations. Figure 4.16a shows the $(P - V)$ diagram for an idealized power development cycle. Segment 1-2 represents the compression cycle, and as shown, while volume decreases, pressure increases as the piston moves up toward the top-dead-center. Segment 2-3 represents the combustion of the fuel-air mixture in the chamber. In the idealized cycle, this combustion of the mixture is assumed to be very quick, so fast that ideally (reality is close to this but not exactly) the volume is not changed, but there is a huge increase in the pressure. Segment 3-4 is the expansion of the gases, the development of power in the engine, when high-pressure gases push down the piston and create the force or moment that rotates the crankshaft. As shown, the pressure decreases while volume increases as the piston moves down. At this point, the exhaust valve is opened and the remaining gas escapes, rejecting the remaining heat. Ideally, this also happens instantaneously, and therefore, at constant volume. The difference between an

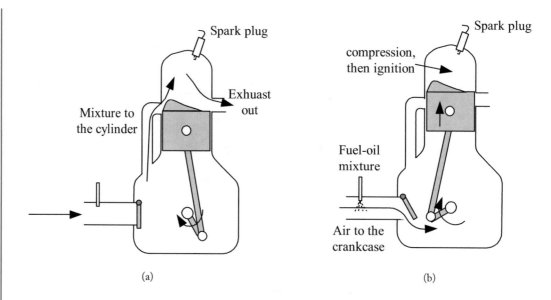

Figure 4.15: The strokes of a 2-stroke engine.

actual 4-stroke engine cycle and the thermodynamic cycle representation is the remaining two strokes of exhaust and intake. In 4-stroke engines the piston moves up and exhaust gases are pushed out (which require a little more energy), and subsequently the piston moves down and fuel-air mixture is pulled in (also requiring a little more energy). However, the ideal thermodynamic cycle does not show these; actual thermodynamic cycle representation in Figure 4.16b includes an additional section to represent the exhaust (segment 4-5) and intake (segment 5-1) strokes. Also notice that in reality, volumes change during ignition and heat rejection. In reality, combustion requires time to complete. Ideally, it is best to start the combustion before top-dead-center and allow it to complete at the same volume that it started. Therefore, in reality, the spark ignition occurs as much as 20° before top-dead-center, and ends as shown in Figure 4.16b at about the same volume.

As mentioned earlier, the ideal thermodynamic representation only represents two of the four strokes of the engine. You may notice that a 2-stroke engine more closely matches this representation, even though the compression and exhaust are not instantaneous as the cycle assumes.

The area surrounded by the four segments in the $P - V$ diagram represents the power developed by the engine at each complete cycle (this is the same as pressure multiplied by volume at each instant. The volume of a cylinder is the area of the base multiplied by its height. In an engine, the volume is the piston area multiplied by the stroke of the piston. Conversely, pressure multiplied by an area is force, and force multiplied by distance is work. Therefore, pressure multiplied by volume is the same as force multiplied by distance, both representing work). An engineer can

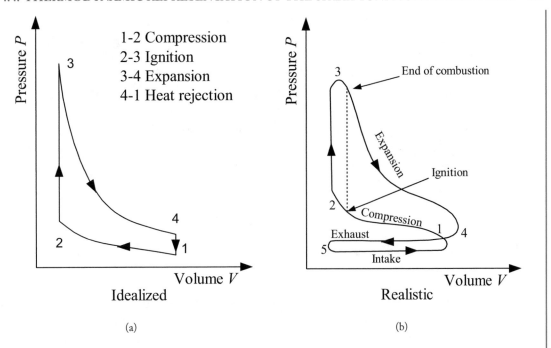

Figure 4.16: Thermodynamic representation of the spark-ignition power cycle.

use this graph to estimate or calculate the power output of the engine. The ratio of the power developed with respect to the chemical energy of the input gasoline determines the efficiency of the engine, and can be estimated from the graph. In reality, the efficiencies are measured under more realistic conditions. It should be mentioned here that fuel injection has increased the efficiency of modern engines. However, there is a large set of desirable characteristics and undesirable consequences that play an important role in the efficiency of engines and cars in general. The shape (aerodynamics) of a car, the desired acceleration and power, the weight of the car, the accessories that are operated by the engine, etc., all affect the efficiency of the car and its MPG rating. High accelerations and high power output means that the engine is very powerful when needed, but in most conditions, its power is excessive and not used, reducing its efficiency significantly. At the same time, we desire to reduce pollution, and therefore add limiting devices and pollution reduction systems to engines that limit their performance and reduce other desirable characteristics. Therefore, like many other engineering decisions, the design of the engine and the chosen size and power characteristics are compromises, based on marketing and engineering considerations.

On a side note, you may have noticed how you were repeatedly asked to *imagine* certain things in order to visualize motions and other happenings in a still picture or figure throughout the book. It is very common in engineering to visualize motions in still pictures and drawings and to see things in one's mind. Whenever we design something new that does not exist we see it in our mind's eye. Some individuals may already be good at this, others learn to do it. In engineering problem solving one needs to visualize things that do not exist and see motions and happenings that are not shown. Whether an engine, a mechanism, a robot, or a thermodynamic cycle, we see much more in a drawing than is shown.

4.5   COMPRESSION-IGNITION DIESEL ENGINE POWER CYCLE

Compression-ignition diesel engines are quite similar to spark-ignition 4-stroke engines. They are predominantly 4-stroke, similarly structured to have intake, compression, power, and exhaust strokes. They also have similar valve operation and construction. The major difference between them is in the way fuel is delivered; in spark-ignition engines, the fuel and air are mixed and compressed before the mixture is ignited by a spark plug ahead of the piston reaching the top-dead-center, whereas in compression-ignition engines only air is compressed and the fuel is injected into it toward the end of the compression stroke. In diesel engines, compression ratios are much higher than in gasoline engines, and therefore, there is no need to use a spark plug to ignite the mixture; it auto-ignites when the fuel is injected into the hot air.

In diesel engines, due to the large compression ratios of 14–20 (even higher in larger systems), the air (and not a fuel-air mixture) is compressed to a high degree, significantly increasing its temperature. Equation (4.3) shows the resulting temperature as air is compressed (temperatures are in Kelvin, $°K = °C +273$ or in Rankin $°R = °F +460$):

$$T_2 = T_1 C_r{}^{0.4}, \tag{4.3}$$

where C_r is the compression ratio and T_1 and T_2 are temperatures before and after compression. Table 4.1 shows the pressure ratios and corresponding temperatures for initial temperature of $20°C = 68°F$ for different compression ratios from Equations (4.2) and (4.3).

When the compression ratio increases, the temperature increases too. For a compression ratio of 8, the approximate temperature will be about $400°C$ ($752°F$), whereas for a compression ratio of 15, it will be $592°C$ ($1100°F$). Since the auto-ignition temperature of diesel fuel is lower than that of gasoline, it can even ignite without a spark at these temperatures. Therefore, instead of a spark plug (and its support system) the fuel is injected into the hot, compressed air.

Table 4.1: Compression ratios and associated pressure ratios and temperatures

Compression ratio C_r	8	11	15	20	25
Pressure ratio P_2/P_1	18	29	44	66	90
Approximate Temperature T_2 $^\circ C (^\circ F)$	400 (752)	490 (914)	592 (1100)	698 (1288)	789 (1450)

The first stroke of a diesel engine is the intake stroke, when the intake valve is opened and air is sucked in. With turbo-charging, the air is pushed into the chamber at slightly higher pressure; this is very helpful at higher elevations where the air pressure is a little lower. As the piston moves toward the top-dead-center in the compression stroke, the air is compressed, increasing its pressure and temperature. Near the top-dead-center, diesel fuel is injected at high pressure into the chamber, and due to the high temperature of the air compared to the auto-ignition temperature of the diesel fuel, it ignites immediately and burns as the piston continues its downward power stroke, creating the torque at the crankshaft. During the fourth stroke, as the piston travels up to the top-dead-center, the exhaust valve is opened, allowing the burnt cases to escape. The valve closes as the piston moves down again while the intake valve is open, repeating the cycle.

Advantages of a diesel cycle include higher efficiencies due to higher compression ratios, lack of an ignition system, and lower cost of diesel fuel (in most places). Diesel engines are usually powerful and are used for trucks, locomotives, marine applications, factories (to generate electricity and run other machines), and even in some power plants. Disadvantages include the lack of availability of diesel fuel as compared with regular gasoline in gas stations (at least in the U.S.), lower power delivery at higher altitudes, more noise, and difficulty starting the engine in cold temperatures. Many diesel engines include a heating element in the combustion chamber for cold-starting the engine. Diesel engines are also more polluting, although they have improved in recent years. However, due to the availability of plenty of air in the mixture, diesel engines produce less CO and more CO_2.

4.6   THERMODYNAMIC REPRESENTATION OF COMPRESSION-IGNITION POWER CYCLE

Similar to the thermodynamic representation of the spark-ignition power cycle, and for the same reasons, we can also represent the compression-ignition cycle (also called *constant pressure combustion cycle*) with both $(T \quad s)$ and $(P \quad V)$ thermodynamic graphs. Figure 4.17a shows the ideal compression-ignition diesel cycle. Segment 1-2 represents the compression of the air in the cylinder as the piston moves up toward the top-dead-center, at which point fuel is injected into the cylinder. Segment 2-3 represents the combustion. The ideal cycle assumes that combustion occurs at constant pressure because fuel continues to burn as the piston moves down. Therefore, the pressure increase due to combustion compensates for the pressure loss due to increase in the volume. Segment 3-4 represents the expansion of gases (development of power), and segment 4-1 represents the rejection of remaining heat or exhaust. Figure 4.17b shows a more realistic diesel cycle where the compression/combustion and expansion is broken into two segments. As in the case of gasoline engines, we can also add the intake and exhaust strokes to the diagram.

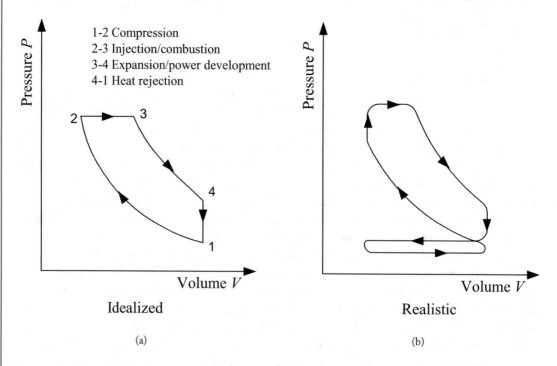

Figure 4.17: Thermodynamic $P - V$ representation of the compression-ignition cycle.

Have you noticed the particular noise that diesel trucks make as they travel downhill at high speeds? This noise is due to braking the truck with the engine instead of powering it with the engine. As we have discussed in previous chapters, kinetic energy of a body is:

$$K = \frac{1}{2}mV^2,$$
(4.4)

where m is the mass and V is the velocity. Trucks are massive, especially when fully loaded. When they travel at high speeds, their kinetic energy is tremendously large. Slowing a truck in a downhill stretch of the highway is almost impossible without damaging the brakes, assuming they even work. To control the speed of a truck in downhill stretches and to slow it down to manageable values, the engine is practically shut down by cutting off fuel to it while still keeping it engaged with the transmission, forcing it to rotate. As a result, the engine acts as a pump, not an engine, requiring work to turn. This work is provided by the kinetic energy of the truck, slowing it down. In other words, in order for the engine to keep turning without fuel, it takes the kinetic energy of the truck and slows it down. To increase this effect, it is possible to alter the opening and closing of the valves and increase the work still required to turn the engine. All these alterations can be studied and designed using the same thermodynamic representations.

It should be mentioned here that we have only looked at three thermodynamics cycles. However, there are many others that relate to other systems, including Stirling and Ericsson cycles, the Carnot cycle, and the Brayton cycle.

4.7   ROTARY (WANKEL) ENGINES

Rotary engines follow a similar thermodynamic cycle, but are mechanically different from reciprocating internal combustion engines. They have an intake, compression, ignition, expansion, and exhaust segments. However, instead of the usual slider-crank mechanism (piston and cylinder, connecting rod, and a crank) rotary engines include an epitrochoid-shaped housing with two openings for intake and exhaust and a three-sided rotor as shown in Figure 4.18. The rotor both rotates and orbits around the fixed geared shaft called the *eccentric-shaft* (e-shaft) with a 1/3 ratio such that for every three rotations of the eccentric shaft, the rotor rotates only once. This forces the three corners of the rotor to always remain in contact with the housing. Spark plugs ignite the compressed fuel-air mixture at the proper time. These engines are simpler, smaller and lighter, and provide a better power-to-weight ratio. However, they are relatively new compared to the reciprocating engines, and therefore, there is less experience available with the design and service aspects of these engines.

As shown in Figure 4.18, at any given time, multiple segments of the cycle happen simultaneously. For example, in Figure 4.18a, the engine is at the end of its intake, in the middle of its power development, and in the exhaust stroke all at the same time. In Figure 4.18b, it is compressing the fuel-air mixture, developing power, and finishing exhaust. In Figure 4.18c the engine is taking in the air mixture, the spark plugs initiate combustion, and exhaust has started. Figure 4.19 shows the rotor of an actual engine in the combustion chamber.

4.8   POWER GENERATION

Although engineers use similar temperature versus entropy $(T - s)$ and pressure versus volume $(P - V)$ diagrams to design and analyze power generation systems (such as in a power plant), we will only discuss the principles of these systems here because in real life, these systems can be complicated and there are too many variations that make each system uniquely different from another, therefore changing the efficiency of the system.

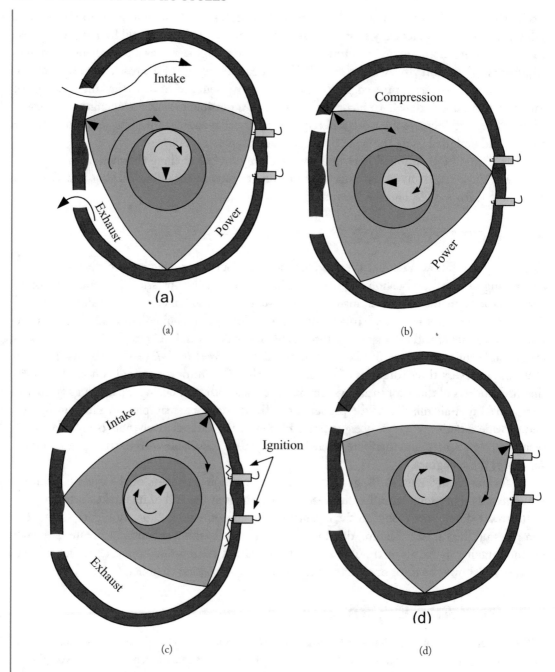

Figure 4.18: Rotary (Wankel) Engine operation.

Figure 4.19: A rotary engine's combustion chamber and rotor in different positions similar to positions shown in Figure 4.18.

As was mentioned earlier, energy is neither created nor destroyed; it is only converted from one form to another. When we speak of power generation in a power plant, we actually mean the conversion of one form of energy such as thermal or hydraulic or chemical energy into electrical. Power generation, among others, includes conversion of energy from coal, gas, or other hydrocarbon and fossil fuels, nuclear, wind, hydraulic, and solar into electrical energy. In each of these systems, a generator is turned at a constant speed to convert the energy into electrical form (see Chapter 6 about generators and motors). The power needed to rotate the rotor of the generator is provided by one of the aforementioned systems.

One of the most common systems used is a steam generator. In these systems, fossil fuel is burned in order to turn water into steam at high pressure and temperature, raising its energy to a very high level. The pressurized and hot steam is then pushed through a steam turbine, causing it to turn. The shaft of the turbine is connected to the generator, and thus, it rotates. The energy needed to boil the water into steam may come from burning coal, gas, other hydrocarbons, nuclear reaction, or similar. Coal is inexpensive and plentiful, but it is very dirty and creates a lot of pollutants, including carbon dioxide. Coal is used all over the world, but in certain countries that use it extensively, the level of air pollution is also very high. Since in recent years gas has become much more available and much cheaper, but burns much more cleanly, many systems have been converted to burn gas.

An important issue here is that, as was discussed earlier, due to the second law of thermodynamics, it is impossible to assume that all the energy of the steam can be converted to electrical power. This will defy the second law. Therefore, at best, the efficiency of a power plant can be as high as low-40s percent. This means that close to 60% of the power in the fuel is wasted as rejected heat.

Another popular system is to turn the generator of a power plant by a jet engine; here, fuel is burned in a jet engine just like the way it is burned in an airplane engine in order to fly it. However, instead of the jet engine pushing through the air to fly the plane, the shaft is connected to the generator, rotating it to generate electricity. The burned air/fuel mixture leaves the system still with high level of energy left in it because it is still very hot and has much kinetic energy (it comes out of the jet engine at a very high speed). Therefore, like most other systems, the efficiency of the jet engine is low and most of the energy is wasted as rejected heat.

An alternative, originally designed decades ago but becoming more popular only recently, is a *combined-cycle*. In a combined cycle, a generator is powered by the jet engine as described earlier. However, the high energy left in the burnt gas is captured by using it to boil water into steam just like the steam power systems. Since more of the energy is captured between the two systems compared to either of them alone, the efficiency of such a system can be more than 60%, a significant increase.

This indicates how engineering principles can be used to make a system better, more efficient, and less polluting. Thermodynamic cycles and the $(T - s)$ and $(P - V)$ diagrams can be used to design and tune the system to its best possible performance level.

4.9   CONCLUSION

As you have probably noticed, the intention of this chapter was not to discuss refrigerators or engines or power generation, but how the study of thermodynamics is necessary in order to know what these cycles are and how they are used by engineers to design and improve these systems. There is, of course, a lot more to thermodynamics than what is discussed here. But hopefully this discussion shows the importance of thermodynamics in our everyday technological lives. There are over a billion cars in the world. Imagine if through thermodynamic studies we could improve their efficiency by a couple of percentage points. Imagine how much energy we would save and how much less pollution we would have to deal with.

4.10   BIBLIOGRAPHY

[1] Gluckstein, M.E and Walcutt, C. "End-Gas Temperature-Pressure Histories and Their Relation to Knock," *Transactions of SAE* 69, 529, 1961. 108

CHAPTER 5

Moments of Inertia

Mass and Area Moments of Inertia, Accelerations,

Inertial Forces, Strengths, and Strains

5.1   INTRODUCTION

If you think of any classic cartoons, it is inevitable that at some point a beloved animal character will slowly crawl on the branch of a tree as it is pursued by its nemesis, bending the branch more and more until it breaks. Have you ever wondered what would happen if the animal had some knowledge of engineering and could calculate how far it could go before the branch would break (knowing about engineering principles makes cartoons even more interesting)? In real life, we can actually predict the strength of the part we are loading and calculate how much load it can safely carry without breaking. Extending the idea of cartoons to real life we can find countless examples where the situation is the same. Simply think of the load on the wings of an airplane. It is actually similar to the situation mentioned earlier. This chapter discusses these relationships and how these ideas are related to each other. Let's start with the following experiment.

Please take a ruler or a piece of wood or similar object and place it between two raised points (perhaps two cups or books) and then press it in the middle as in Figure 5.1a. You will notice that bends. A larger force exerted by you will cause the ruler to bend more.

Now turn the ruler 90° on its side and repeat as in Figure 5.1b. You will notice that the ruler, under the same force, does not bend at all (or bends very little). So why is it that although it is the same object, with the same dimensions, the same mass or weight, and the same strength, that in one orientation it bends more easily than in another orientation?

The reason is that the moment of inertia of the object, in this case the *area moment of inertia*, is different between the two orientations. Although everything else is the same, the area moments of inertia are *not*. The same is true when we deal with the motion of objects, where *mass moment of inertia* is a factor. In this chapter we will examine how moments of inertia affect the behavior of objects, both as they relate to static (not moving) and dynamic (moving) situations.

It should be mentioned here that area moment of inertia is not an accurate description of this entity, but it is a name which we commonly use. A better name would be *second moment of the area*.

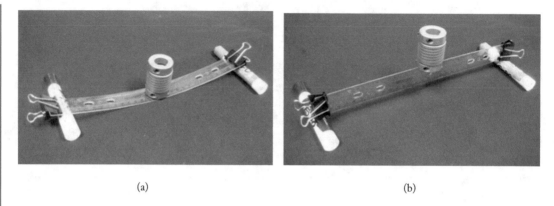

(a)   (b)

Figure 5.1: How much a ruler bends under the same load depends on its area moment of inertia (second moment of the area) in that orientation.

In fact, we can define both the first and second moments of an area. The first moment of area is used to calculate the center of the area, and although this has many applications, we will not discuss it here. We will first discuss the area moment of inertia; mass moment of inertia will be discussed second.

5.2   SECOND MOMENT OF THE AREA (AREA MOMENT OF INERTIA)

Second Moment of Area (or Area Moment of Inertia) is a representation of the dimensions of an area and its distribution (thin, tall, round, square, hollow). Among others, second moment of the area is a measure of how much a body resists bending under a force or resists rotation under a torque (such as in the rotation of one end of a shaft relative to the other end when twisted).

Let's consider a simple bar with a rectangular cross section as shown in Figure 5.2. Although it is very easy to derive the second moment of a rectangular area by integration, we will skip this derivation here. Let it suffice to say that the second moment of a rectangular area about the x-axis is:

$$I_x = \frac{1}{12}bh^3, \tag{5.1}$$

where b is the length of the base of the rectangle and h is its height. Notice that the second moment of the area is independent of the length of the bar.

In order to get a feel for the numerical value of the second moment of the area of a rectangular beam we need to look at a few examples with real numbers. Please stay with the numerical examples as they clarify the point much better than a simple equation. So let's assume that the base of the beam is one inch and the height is 4 in. The second moment of the area of the beam

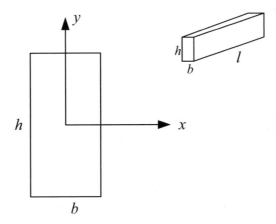

Figure 5.2: The rectangular cross section of a beam.

will be:

$$I_x = \frac{1}{12}(1)(4)^3 = 5.33 \text{ in}^4.$$

Notice that the unit for the second moment of the area is in^4 (or cm^4, etc.). Also notice that this is a measure of the area and its distribution, meaning the size of the area and its relative width and height, but that it has nothing to do with what kind of material it is or how strong it is. Note also that for this example, the area of the cross section is $1 \times 4 = 4$ in^2.

Now let's do the same, but this time we will turn the beam on its side such that the base will be 4 in while the height is 1 in. Notice that the area is still the same $4 \times 1 = 4$ in^2, but this time, the area moment of inertia will be:

$$I_x = \frac{1}{12}(4)(1)^3 = 0.33 \text{ in}^4.$$

As you notice, even though the beam is exactly the same, with the same dimensions and the same area, its second moment of area has changed significantly, in this particular example a ratio of 5.33/0.33 or more than 16/1, all because we simply turned it around.

To better understand this, let's now consider a beam with a square cross section of 2×2 in. Here too, the area is the same $2 \times 2 = 4$ in^2 as before, but the area moment of inertia is:

$$I_x = \frac{1}{12}(2)(2)^3 = 1.33 \text{ in}^4,$$

and the ratio, compared to the previous case, is 5.33/1.33 or 4/1. Once again, factors influencing the magnitude of the second moment of an area are both the actual dimensions of the area and their distribution.

We can also define the area moment of inertia of the same cross section about the y-axis (we will see the application of this shortly). In this case, as in Figure 5.2, relative to the y-axis the height h will be the base and the base b will act as the height, and therefore:

$$I_y = \frac{1}{12}hb^3.$$

Substituting the same dimensions as before, we will get the second moment of area about the y-axis as 0.33 in^4. Notice that the second moment of area about the x-axis for our first case when $b = 1, h = 4$ is the same as the second moment of area about the y-axis for the second case when $b = 4, h = 1$.

If we calculate the approximate second moments of area of the ruler of Figure 5.1 where the ruler is 0.075 in thick and 1.175 in wide, we get the following:

For Figure 5.2a: $I_x = \frac{1}{12}(1.175)(0.075)^4 = 0.000003$ in^4

For Figure 5.2b: $I_x = \frac{1}{12}(0.075)(1.175)^4 = 0.0119.$

The ratio is 0.0119/0.000003 =3,840. This means it will take 3,840 times as much force to bend the ruler of Figure 5.1a the same amount as Figure 5.1b, an amazing difference (meaning that most probably, the ruler will break before it bends).

Before we continue our discussion of the second moment of area, let's see how it is used in calculating the deflection of the beam under a load as well as its stresses.

5.3   DEFLECTIONS OF A BEAM

Deflection relates to how much a beam bends; it is usually calculated at its maximum, in this case, in the middle of the beam. The maximum deflection for a simple beam, supported at the two ends (like the ruler of Figure 5.1) and a single force in the middle can be calculated by:

$$y_{\max} = -\frac{FL^3}{48EI},\tag{5.2}$$

where $y_{\max}$ is the maximum deflection at the center of the beam, F is the load, L is the length of the beam, E is called *modulus of elasticity*, a material property (which we will discuss later), and I is the second moment of the cross sectional area of the beam, as shown in Figure 5.3a. The negative sign indicates that the beam bends down (below the reference frame x-axis). Since the second moment of the area I is in the denominator, as it gets larger the deflection decreases. As you can see, when the ruler of Figure 5.1 is turned 90°, the only thing that changes in this equation is the second moment of the area of its cross section; otherwise, the load, the modulus of elasticity, and the length remain the same. If the second moment of the area is 3,840 times as large, the deflection will be 3,840 time smaller compared to the first case, and this is exactly what we see. This fact is used extensively in the design of structures and machine elements in order to limit or increase deflections as necessary. For example, a roof beam should not deflect much, and therefore, the beam is laid in the direction of the maximum second moment of the area, whereas in the leaf spring of Figure 5.4, used at the rear axle of a truck, the beam (each leaf of the spring) is laid on its base to increase the deflections, therefore acting as a spring.

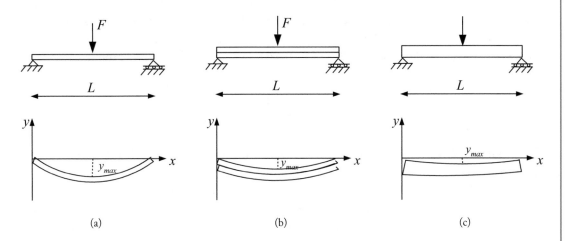

Figure 5.3: The deflection of a beam under a load at its center.

Figure 5.4: A leaf spring used in a truck.

The second moment of the area used in Equation (5.2) is I_x. So when do we use I_x and when I_y? It depends on about what axis the beam bends. For example, if the beam of Figure 5.2

is loaded with a force in the vertical direction, causing it to bend about the x-axis, we use I_x. If the beam were loaded with a horizontal force causing it to bend about the y-axis, we would use I_y.

Now assume that instead of one ruler we would use two of them on top of each other as in Figure 5.3b. In this case, since they both bend under the same load, we can assume that each one will carry almost $1/2$ of the load, and therefore the deflection will be $1/2$ of the first case, or similarly, that the total second moment of the area is twice as much for two of them, and therefore:

$$y_{\text{max}} = -\frac{FL^3}{48E(2I)}.$$

Notice that this means that the two rulers *slide* over each other as they bend. This can be likened to bending a telephone book. As you bend the book, all the pages bend together and slide over each other, but they all maintain their original lengths.

Now let's assume that instead we use a similar ruler or beam, but twice as thick. In this case, the total amount of material would be the same as using two thinner rulers or beams, and the overall dimensions would be similar. However, the second moments of the area are different. Whereas with two rulers, the total second moment of the area is:

$$I_{total} = 2 \times \frac{1}{12}bh^3 = \frac{1}{6}bh^3,$$

the total second moment of the area for a beam twice as thick (with its height equal to twice the height of the original beam) is:

$$I_{total} = \frac{1}{12}b(2h)^3 = \frac{1}{12}b(8h^3) = \frac{2}{3}bh^3,$$

which is four times as large as the first case with a deflection four times smaller. Notice that unlike the first case where the two beams slide over each other, in this case there are no separate layers to slide over each other; the beam bends as one piece. This small difference is the reason for the different magnitudes of deflection (and as we will see later, stresses). It is as if you held all the pages of the telephone book together while trying to bend it, if they were all glued; it would strongly resist bending. The sliding of the layers of the beam over each other is called *shear*. When the layers slide over each other and consequently there is no resistance between them, there is no shear force; when they are prevented from freely sliding over each other, there is a shear force between the layers, and consequently, the book does not bend. Figure 5.5 shows this difference between a telephone book whose pages are prevented from sliding by two large paper clips versus free sliding of the pages. This also explains why a cardboard is much stronger (stiffer) than individual papers with the same thickness. The paper layers in a cardboard are glued together, therefore preventing them from sliding over each other. Plywood is made of thin layers of wood, glued together. They are strong and resist bending too. However, plywood is also used in the

Figure 5.5: When the pages of a telephone book are prevented from sliding over each other, it does not bend as much due to the differences in second moments of the area.

manufacture of bent surfaces such as in modern furniture. In this case, the thin layers are first bent to shape, then glued together. Therefore, they maintain their shape.

So why does the sliding of layers over each other matter? To understand this, let's once again look at the cross section of the beam, in this case a rectangle. As you see in Figure 5.6, the centerline of the cross section is called *neutral axis* which is through the center of the area (called *centroid*). For symmetrical cross sections such as a rectangle or a circle, the neutral axis is in the middle. At the plane of the neutral axis the length of the beam does *not* change during bending. This means that as the beam bends, its length remains the same at the neutral axis, while all other layers change length. In the case of a beam loaded from above as shown in Figure 5.6, all layers of the beam above the neutral axis must shorten while all layers of the material below the neutral axis must lengthen. The farther away a layer is from the neutral axis, the larger the increase or decrease in its length. Now imagine how much larger the increase and decrease will be when the cross section increases in height. This is why the bending of the beam decreases significantly as the height of the beam increases, which is reflected as h^3 in the second moment of the area equation.

Now compare this with doubling the number of beams instead of doubling the height. In the case of two beams, the lengths of the layers of each beam increase or decrease independently based on the beam's height, while each one slides over the other beam. (Think about the layers of the upper beam below its neutral axis which lengthen, while the layers of the lower beam above its neutral axis which shorten. At the interface, one is shortened, one is lengthened. Consequently, they slide over each other.) Because the height of each individual beam is less, the increase or decrease in length of the layers is much less. Once again, think about the telephone book and how the lengths of the pages must increase or decrease when glued together versus the pages sliding over each other when not glued. However, when the height of the beam is doubled, there

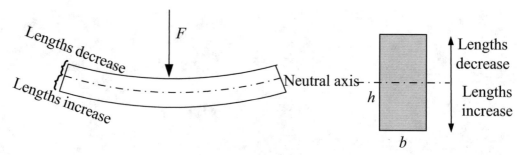

Figure 5.6: The neutral axis of a rectangular cross section.

is no sliding of the layers; the farther away a layer is from the neutral axis, the larger its shortening or lengthening.

Obviously, it is possible to have cross sections other than a rectangle. Examples include circular (such as a shaft), hollow circular (such as a tube), hollow rectangular, I-beams, C-channels, L-shaped angles, and many more. There are either formulae for calculating the second moments of area for these shapes, or they can be found in tables.

The second moment of a circular area is:

$$I_x = I_y = \frac{1}{4}\pi r^4, \tag{5.3}$$

where r is the radius of the circular cross section. Notice that due to the symmetry of a circular area, the second moments about the x-axis and y-axis are the same.

The second moment of the area for a tube can easily be calculated by subtracting the second moment of the inner area (treated as missing material or as a negative moment) from the second moment of the total area as:

$$I_x = \frac{1}{4}\pi r_o^4 - \frac{1}{4}\pi r_i^4, \tag{5.4}$$

where r_o and r_i are the outer and inner radii of the tube as shown in Figure 5.7a. The same can be done for a hollow rectangular tube or other shapes. Similarly, we can add second moments of the area together for shapes that are combinations of elements with which we are already familiar. For example, suppose that two rectangular beams of the same size are placed next to each other as shown in Figure 5.7b. The total second moment of the area about the x-axis for both will be the summation of the moments, or:

$$I_x = \frac{1}{12}bh^3 + \frac{1}{12}bh^3 = \frac{1}{6}bh^3. \tag{5.5}$$

If you have ever worked with electrical wires you have probably noticed that multi-strand wires are much easier to bend than single-strand wires of the same gauge (thickness). The reason

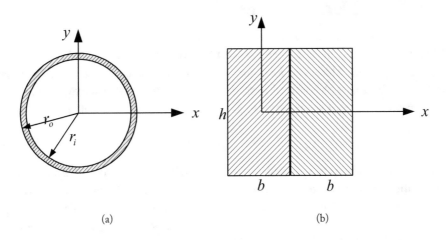

(a)   (b)

Figure 5.7: Second moments of the area for combined areas.

is that multi-strand wires consist of many thinner wires that can slide over each other. The total second moment of the area is the summation of the moments of each strand. However, the second moment of the area for the thicker single-strand wire is much bigger than the second moment of the area for the multi-strand wire, and consequently, it is stiffer. The same is also true for a steel cable versus a steel bar of the same diameter. Cables are much easier to bend than bars because the second moment of the area for a bar is much bigger too. The strands of the cable can slide over each other; the layers of the bar cannot.

Tree branches are the same. Thicker branches have a larger diameter which increases the area moment of inertia, reducing deflection under the force of winds and the weight of its fruit, other branches, animals, and leaves. Being as smart as it is, nature provides adequate strength as necessary. Because the loads decrease as we get closer to the top of the tree or to the tip of each branch, its thickness reduces as well. This reduces the weight and optimizes the design; there is basically enough material to take the load as needed.

Second moments of the area for most common standard building beams such as I-beams are available in manufacturers' tables where engineers readily find them. However, for shapes that are not included in tables or are not common, we can easily calculate the second moments of area, some by mathematical integration, others by combining formulae used for common shapes. To understand this, which also helps in further understanding the idea of the second moment of area, let's consider the *Parallel Axis Theorem*.

5.4 PARALLEL AXIS THEOREM

As you may have noticed, we calculated the second moment of the area about the neutral axis (we placed the origin of the reference axes at the center and calculated the moments relative to the x-axis and y-axis). However, for many different reasons (which will become clear shortly) we may need to calculate the second moment about other axes away from the neutral axis. The second moment of the area about another axis x', parallel to the neutral axis, can be found from:

$$I_{x'} = I_x + Ad^2,\tag{5.6}$$

where A is the area and d is the distance between the two axes. In other words, the second moment of the area about x' is equal to the second moment about an axis through the centroid plus the area multiplied by the square of the distance between the two axes. This is called parallel axis theorem. For example, the second moment of the area of a rectangle about the bottom of the rectangle instead of its centerline (Figure 5.8) is:

$$I_{x'} = I_x + Ad^2 = \frac{1}{12}bh^3 + (bh)\left(\frac{h}{2}\right)^2 = \frac{1}{12}bh^3 + \frac{1}{4}bh^3 = \frac{1}{3}bh^3.$$

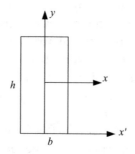

Figure 5.8: Parallel axis theorem.

Now let's see where this can be used. Imagine that we model an I-beam (Figure 5.9b), a very common structural beam element whose second moment of area is often needed for stress and deflection calculations, as three rectangular-shaped areas attached to each other as shown in Figure 5.9a. The vertical portion is called a *web* and the horizontal portions are called *flanges*. In this case, the total second moment of the area about the neutral axis x is the summation of the second moments of each of the three areas, all about the neutral axis x. The second moment of the web can easily be calculated by Equation (5.1). However, the second moments of the flanges must also be calculated about the same x-axis that was used for the web, which is a distance of d away from each flange. Therefore, we will need to use the parallel axis theorem Equation (5.6) to calculate the contribution of the flanges to the total second moment. The total second moment of area for the I-beam about the x-axis is:

$$I_{total_x} = I_{web_x} + 2I_{flange_x}.$$

The second moment of the web about the x-axis is:

$$I_{web_x} = \frac{1}{12}(t)(h^3).$$

The second moment of each flange about its *own* axis x' and x'' is:

$$I_{flange_{x'}} = I_{flange_{x''}} = \frac{1}{12}(b)(t^3).$$

But since we need the second moment about the x-axis, we use parallel axis theorem and get:

$$I_{flange_x} = I_{flange_{x'}} + Ad^2 = \frac{1}{12}(b)(t^3) + (bt)(d^2)$$

$$I_{flange_x} = \frac{1}{12}bt^3 + btd^2.$$

Therefore, the total second moment for the I-beam is:

$$I_{total_x} = \frac{1}{12}th^3 + 2\left(\frac{1}{12}bt^3 + btd^2\right).$$

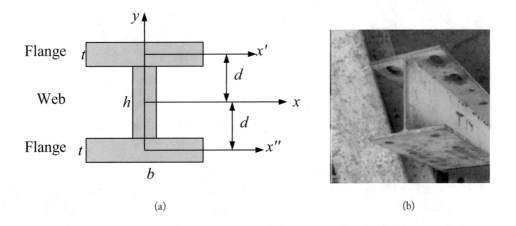

Figure 5.9: (a) A simplified model of an I-beam, (b) An actual I-beam.

To see the significance of this let's assume we make an I-beam out of three pieces similar to our previous example, 1×4 in. If the three were laid next to each other as in Figure 5.10a, the total second moment of the area would be the summation of their individual moments about the x-axis:

$$I_{total} = I_{web} + 2I_{flange}$$

$$= \frac{1}{12}(1)(4)^3 + 2\left(\frac{1}{12}(4)(1)^3\right) = 6 \text{ in}^4.$$

However, if they were assembled (and glued/welded together) into an I-beam as in Figure 5.10b, the second moment of the area would be:

$$I_{total} = \frac{1}{12}th^3 + 2\left(\frac{1}{12}bt^3 + btd^2\right)$$

$$= \frac{1}{12}(1)(4)^3 + 2\left(\frac{1}{12}(4)(1)^3 + (4)(1)(2.5)^2\right) = 56 \text{ in}^4.$$

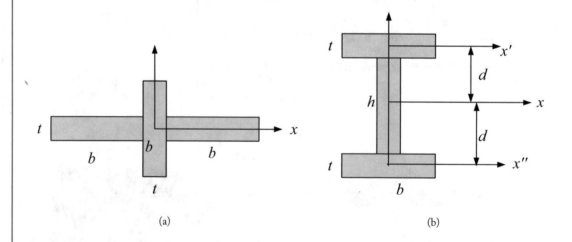

Figure 5.10: An I-beam versus its constituent parts next to each other makes a huge difference in the total second moment of the area.

This is over nine times as large. The fact that the flanges are at a distance away from the x-axis significantly adds to the total second moment as compared to the flanges on the x-axis. This is why the distribution of the material, and not the total area, is important in how much load a beam carries or how much it deflects under the load. This example shows the importance of the shape of the beam and how much load the same material carries in a structure or a machine. Now suppose that the same amount of material is used either as a flat sheet or as an I-beam by cutting it into three strips and gluing the pieces together in the shape of an I-beam (Figure 5.11). Even though they are the same amount of area (same material), the I-beam will carry a much larger

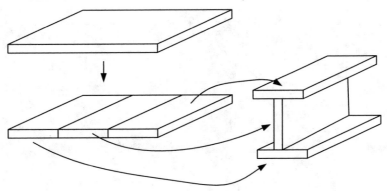

Figure 5.11: A strip of material versus cutting and gluing it into an I-beam. The I-beam carries significantly more load than the strip.

load. Engineers can design structural elements that are much more efficient with less material because they use these engineering principles in their designs.

Another major example of where the second moment of the area is increased by distributing the material farther away from the neutral axis is the use of a truss. Because the distance of the element of the truss from its neutral axis is increased, its moment of inertia is also increased significantly, enabling it to carry larger loads, especially at larger spans. Figure 5.12 shows an example of a truss used as the main load-carrying element in a ceiling. Look for it in a bridge next time you see one.

Figure 5.13 shows a common dish rack. Can you tell why the body is designed this way? In addition to their effect on the shape of the rack, the two sets of semi-circular welded horizontal members create a much larger second moment of area than if they were added together as a thicker rod, were laid next to each other, or were free to slide over each other.

Figure 5.14 shows two corrugated pieces of cardboard. Looking at their cross sections and the differences between their construction, can you tell why the one in Figure 5.14a rolls easily in one direction for wrapping purposes (but not in the perpendicular direction), whereas the corrugated cardboard of Figure 5.14b is stiff in all directions?

5.5 POLAR MOMENT OF INERTIA (POLAR MOMENT OF THE AREA)

So far we discussed the role of the second moment of the area in bending. Now imagine a similar situation, but here we intend to twist a bar by applying a moment or torque to one end as shown in Figure 5.15. This twisting of the bar is called *torsion*. As in bending, when a bar is twisted, one end of the bar rotates relative to the other end. This twisting of the bar is called *angular deflection*.

Figure 5.12: A truss used as a load-carrying element in a ceiling. The second moment of the area of the truss is significantly larger due to the way the elements are distributed farther away from the neutral axis.

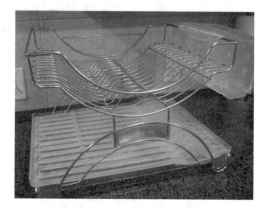

Figure 5.13: The two sets of semi-circular horizontal members of the dish rack increase the second moment of the area, decreasing its deflections.

In torsion, we use the *polar moment of the area* (polar moment of inertia), J, which for a round shape is:

$$J = \frac{1}{2}\pi r^4.$$

(5.7)

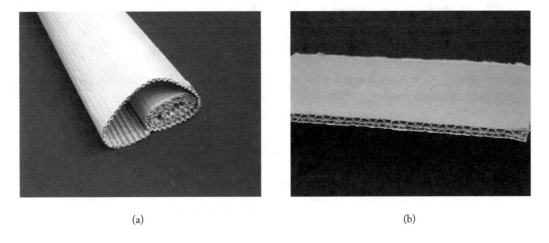

(a)   (b)

Figure 5.14: Corrugated cardboard is stiff due to its increased area moment of inertia.

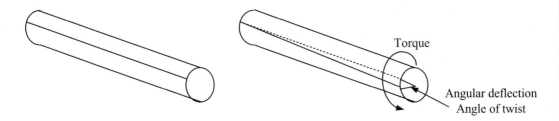

Figure 5.15: Torsion of a bar.

Notice that this is twice as large as I_x (Equation (5.3)), which makes it equal to $I_x + I_y$ (for a symmetrical cross section). The polar moment of the area can be similarly calculated for other shapes, including with the use of the parallel axis theorem.

Similarly, we can also define *modulus of rigidity* G, a material property similar to the modulus of elasticity E that was used in calculating deflections in Equation (5.2). In torsion, the angular deflection (angle of twist) is:

$$\phi = \frac{TL}{JG},$$
(5.8)

where ϕ is the angular deflection, T is the applied torque, and L is the length of the bar. When torque or the length of the bar increase, the angular deflection increases as well. However, as the polar moment of the area increases, the angular deflection decreases. Similar to bending, the polar moment of the area directly affects the twisting of the bar.

The polar moment of the area for a hollow bar is:

$$J = \frac{1}{2}\pi r_o^4 - \frac{1}{2}\pi r_i^4, \tag{5.9}$$

where r_o and r_i are the outer and inner radii of the bar. This is the polar moment of the larger area, minus the polar moment of the missing (hollow) area. Now imagine a solid shaft with a radius of 0.5 in. The polar moment of the area will be:

$$J = \frac{1}{2}\pi r_o^4 = \frac{1}{2}\pi(0.5)^4 = 0.098 \text{ in}^4.$$

The cross sectional area of the shaft will be:

$$A = \pi r^2 = \pi(0.5)^2 = 0.785 \text{ in}^2.$$

Now imagine that we use the same amount of material (same cross sectional area), but we make the shaft hollow. In this case, the outer diameter of the shaft will have to increase to accommodate the hole and still have the same area. There are countless different choices available for the inner and outer diameters to achieve the same area. Let's choose the outer diameter of 0.75 in. In that case, the inner diameter will be:

$$0.785 = \pi(0.75)^2 - \pi(r_i)^2$$
$$r_i = 0.56.$$

Therefore, the shaft will be a hollow tube with 0.56 and 0.75 inner and outer radii as shown in Figure 5.16. The polar moment of the area with the new dimensions will change to:

$$J = \frac{1}{2}\pi r_o^4 - \frac{1}{2}\pi r_i^4 = \frac{1}{2}\pi(0.75)^4 - \frac{1}{2}\pi(0.56)^4$$
$$J = 0.343.$$

Notice how much larger the polar moment of the area is although the same amount of material has been used. The new polar moment of the area is $0.343/0.098 = 3.5$ times as large. As long as we do not make the new shaft's wall thickness so small that it will collapse under the load, increasing diameter also increases the polar moment of the area. Obviously, this is much more efficient in material use.

An example of this is the driveshaft of an automobile. When the engine of a car is in the front but the car is rear-wheel driven (examples include many older cars, some larger cars, and most trucks), a driveshaft connects the transmission (in the front) to the differential (in the back) as shown in Figure 5.17. In order to increase the efficiency of the system and lower the weight and the cost of the car, the shaft is hollow.

So far we only discussed the role of second moment of the area (and polar moment of the area) in deflections. Actually, although in certain applications deflection calculations might be the

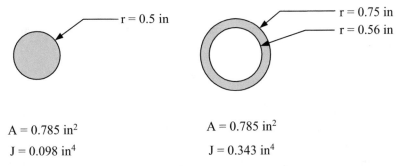

Figure 5.16: The polar moment of the area increases significantly as the shaft is made hollow but with the same area.

Figure 5.17: The driveshaft of a car connects the transmission to the differential.

primary concern, in most cases stress calculations are even more important because stress calculations determine whether or not a structural or mechanical element can carry the load to which it is subjected. Before we discuss this issue, let's first look at the material strength characteristics and see how they are related to moments of the area and mechanical stress.

5.6   STRENGTH OF MATERIALS: STRESS, STRAIN, AND MODULUS OF ELASTICITY

If you attach a weight to a spring it will stretch (elongate). For larger weights, the stretch will be larger. If you plot the weights versus elongations, you will notice that for the most part, the relationship is linear. This means that for example, if the weight is doubled, the elongation will be doubled too. Therefore, we can define a relationship between the weight (which is a force) and the elongation as:

$$k = \frac{F}{d},$$
(5.10)

where F is the force (or weight) in lb or N (Newton, a unit of force in SI system of units), d is the elongation in inches or meters, and k is the *spring constant* in lb/in or N/m. k is a measure of the stiffness of the spring; the larger the stiffness, the harder it is to stretch or compress the spring. Now imagine that you continue to add to the weight until the spring stretches to its fullest. At that point, the spring does not stretch as freely as before. Therefore, the relationship between the force and deflection changes at this point and it becomes much stiffer, a non-linear relationship. Figure 5.18 shows a simplified depiction of this behavior.

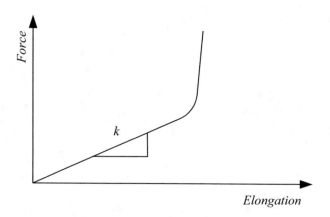

Figure 5.18: The spring stiffness is the ratio of applied force and the resulting elongation.

A similar thing happens to a metal bar. Figure 5.19 shows a typical bar that is used to study the characteristics of many materials, including metals. The bar is placed in a machine that pulls the two ends, applies a force to the bar, and measures the elongation of the bar under the load. In this example the bar was pulled until it broke. Note how the area that broke was reduced in diameter before breaking. As it may be clear to you, the bar's elongation is influenced by how thick it is; the thicker the bar, the smaller the elongations for the same force. Therefore, to measure the strength of the material without the influence of its size, the force is normally divided by the area. This is called *stress*. It may also be clear that the longer the bar is, the larger the total elongation will be (think of a short rubber band and a long one; the long rubber band stretches more than the short one). Therefore, in order to eliminate the effect of length and measure only the material property, the elongation is divided by the length of the bar. This is called *strain*. Consequently, we can study the relationship between stress and strain. This way, the relationship is about the behavior of the material without regard to its thickness or length. Therefore:

$$\sigma = \frac{F}{A},$$ (5.11)

where σ (read *sigma*) is the stress, F is the applied force, and A is the cross sectional area of the bar, and:

$$\varepsilon = \frac{l'}{l},$$ (5.12)

where ε (read *epsilon*) is the strain, l is the original length, and l' is the elongation.

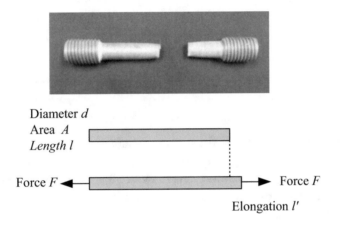

Figure 5.19: A typical material testing specimen.

Calculation of the stress in any structural or machine element is crucial in ensuring that the structure or the machine will be able to take the load to which it is subjected. If stresses exceed the strength of the material, it will fail. Stress calculations are among the most important activities that a design engineer might perform. At other times, not just the stresses but also the deflections are considered because excessive deflections may also cause failure. Therefore, it is not just the calculation of stresses and strains, but also the understanding of the behavior and strengths of the material used that are extremely important in engineering design and analysis.

Figure 5.20 shows the relationship between stress and strain for steel. As you see, when a force is applied to steel, it stretches; when the force is removed, it returns to its original length. Therefore, a steel bar acts just like a spring, albeit it is much stiffer. This is an extremely important characteristic that is used in many engineering calculations to ensure that a part can carry the load to which it is subjected. As shown in Figure 5.20, to a certain point, the steel bar will elongate proportionally to the applied force, at which point its behavior changes. This is called *proportional stress limit S_p* and it is a very important characteristic. The ratio of stress over strain within this limit (the slope of the line) is also a very important characteristic of materials, called the *modulus* *of elasticity, E:*

$$E = \frac{\sigma}{\varepsilon}.$$

(5.13)

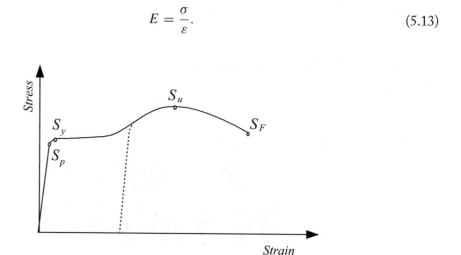

Figure 5.20: The relationship between the stress and strain of steel.

Modulus of elasticity for common steel is about 30×10^6 psi or 200 GPa (Giga-Pascal) in SI units. This means that although a steel bar behaves like a spring, it requires 30 million lbs of force per square-inch of the area to elongate it 1 in/in.

Up until the proportional limit, steel behaves linearly. For small additional forces, the elongation is not linear, but when the force is removed, the bar still returns completely to its original length without any permanent change in its length. This limit is called *elastic limit* or *elastic strength* or *yield strength*, S_y. However, if the force is increased beyond this limit, the bar will permanently elongate, although when the force is removed the bar shortens an amount representing the elastic elongation. So for example, if a machine part is subjected to a force large enough to take it beyond the elastic limit, it will permanently change (and this is why it is called yield strength, because at this point the material yields to a new shape or length). This change is called *plastic deformation*. When the load is removed, it shortens an amount equal to the elastic deformation, but with a permanent elongation that no longer goes to zero. In many situations in design, this is considered a failure of the part, even if it has not broken. Imagine a part of an engine permanently elongating while rotating; the engine will no longer function properly even if no parts are broken. However, most parts are designed with a safety factor to ensure we do not reach the yield strength. And this is why we dare load a car with large loads, but still do not expect it to plastically change forever. We know it deforms under the additional load, but since it is elastic, as soon as the load is removed, it returns to its original shape.

If the force is increased further, even for relatively smaller amounts, the steel bar will elongate in much larger amounts until it eventually approaches the maximum stress it can take, called *ultimate strength*, S_u. At that point, the cross section of the area becomes smaller (because it plastically yields) and the material breaks and fails at S_F.

Other types of steel behave somewhat differently. For example, a piece of high-carbon steel is much stronger, but also very brittle, and therefore, does not elongate as much although it can take higher loads. Therefore, the stress-strain graph representing it, as well as its proportional, yield, and ultimate strengths and its modulus of elasticity will be different. Similarly, other metals (aluminum, brass, copper, stainless steel, and other alloys) all behave a little differently, but follow similar patterns. Additionally, other materials such as glass, concrete, wood, and plastics can also be characterized similarly even though the numbers and the patterns of behavior may be different. For example, glass is a very hard but brittle material. Therefore, it does not yield much, and if subjected to bending it suddenly breaks before any permanent elongation or yielding has occurred.

As shown in Figure 5.20, when a part is subjected to loads beyond its yield strength, it permanently deforms, although the elastic deformation is recovered when the load is removed. However, what happens here is that if the same part were to be subjected to a new load, the load would have to be *larger* than before in order to permanently deform the part again (shown as the dotted line). This is because, as you may notice in Figure 5.20, the portion of the graph between the yield strength and ultimate strength has a small upward slope, and therefore, every time the load must be larger to have the same effect. This means that the material actually becomes harder and stiffer every time it is loaded beyond the yield strength. This is called cold-working, and is a common method of strengthening parts. For example, cold-rolled steel is stronger than hot-rolled steel because if it is heated, the material is softer and it does not require as much force to yield it.

It is interesting to note that human nature is somewhat similar. People who never work hard or never endure hardships behave differently than people who experience difficulties and hardships and learn from these experiences. A broken toe, an illness, lost belongings, failures, and social difficulties all contribute to our resilience. Every experience that involves some hardship beyond our "yield limit" will make us tougher. We even have a name for people who have never had hardships. We call them spoiled. And just like metals, where if the loads become too large for the material it will break and fail, we hope that the hardships to which humans are exposed will not be beyond their capability, causing complete failure. Otherwise, experiences with hardships are good for us; they make us stronger.

5.7   ROLE OF MOMENTS OF THE AREA IN STRESS CALCULATIONS

Now that we have learned about stresses, we can go back to the previous discussion about moments of the area.

In Sections 5.3 and 5.5 we discussed the linear deflection of a beam in bending and angular deflection of a bar in torsion and saw how we can calculate these deflections for simple elements. Similarly, we can calculate the stresses in bending and torsion (and of course in more complicated loading situations that we will not discuss here). For a bending beam as in Figures 5.3 and 5.5, the maximum stress (which happens to be in the middle of the beam) is:

$$\sigma = \frac{MC}{I}, \tag{5.14}$$

where σ (read sigma) is the stress, M is the moment, C is the distance to the top or bottom of the beam from the neutral axis (for maximum stress), and I is the second moment of the area. The second moment of the area also plays a fundamental role in the calculation of stresses as it does for deflections. The larger the second moment of the area is, the smaller the maximum stress will be. Also notice that if we let $C = 0$, indicating a distance of zero from the neutral axis, Equation (5.14) shows that the stress on the neutral axis will be zero (and consequently, there is no deflection either); it linearly increases as we move away from the neutral axis to the top or bottom.

Similarly, for torsion, the maximum (shear) stress in the bar of Figure 5.15 is:

$$\tau = \frac{Tr}{J}, \tag{5.15}$$

where τ (read tau) is the shear stress, T is the applied torque, r is the radius of the bar, and J is the polar moment of the area as discussed in Section 5.7. Once again, polar moment of the area is a fundamental element in the calculation of stresses as well as deflections. The larger the polar moment of the area is, the smaller the shear stress will be. This also shows that the stress at the center of the bar (where $r = 0$) is zero, increasing as we get closer to the outer edge. In fact, the material closer to the center is almost wasted; it carries little load (because stresses are low). This is another good reason to use a hollow shaft rather than a solid one. The same material spread out into a hollow shaft will have a larger polar moment of inertia and will also save on wasted low-stress material.

The diameter of tree branches becomes smaller closer to the tip compared to the base as shown in Figure 5.21. Since the load on the branch is smaller closer to the tip, the diameter and the moment of inertia of the branch are smaller, resulting in less weight and increasing the tree's efficiency.

The second moment of the area and the polar moment of the area are very important concepts in engineering. Understanding the role of moments of the area in this process is a fundamental requirement for engineers.

Figure 5.21: Tree branches become smaller closer to the tip because the load on the branch is smaller too.

5.8   MASS MOMENT OF INERTIA

As area moments of inertia (including the polar moment of the area) are a representation of the distribution of the area, mass moments of inertia are representations of how the mass is distributed. So, even though two different parts may have the same total mass, depending on their shape, their mass moments might be very different. Similarly, as the area moments of inertia directly impact how the material reacts under the influence of external loads and how large the stresses and strains are, mass moments of inertia directly affect the way the mass reacts to accelerations, causing it to move differently depending on not just the mass, but its distribution too. This has a direct effect on our daily lives and the way things move and react as we work with them.

So let's first talk about the context in which mass moments of inertia play a role before we learn what they are and how to calculate them for simple cases.

In Chapter 3 we had a discussion about linear accelerations and how mass reacts to accelerations (please review if necessary). As mentioned there, imagine that you are sitting in a car, accelerating forward. You will notice that you are pushed back against the seat. In this case, since the acceleration vector is forward (causing the car to speed up in the forward direction), the mass of your body reacts to this acceleration; due to its inertia (sluggishness), the body tends to stay in the condition it is in and not change, and therefore, reacts to a push forward by resisting it. The same is true in other conditions. For example, if a body is moving at a constant speed it tends to remain at that speed and reacts to speeding up or slowing down. Therefore, when braking and consequently having a backward acceleration, the body tends to move forward to react to it unless it is restrained. In fact, if this deceleration happens very quickly (such as in an accident when the slow-down is extremely quick, creating a huge backward acceleration), the body may spring forward enough to hit the front windshield. This is why we have seat belts and airbags to restraint the body and keep it from hitting the windshield. Please see Sections 3.2.4 and 3.4 for additional discussions.

The discussion above is about the relationship between a force, linear acceleration, and mass. A similar relationship exists in angular motion. In this case, the relationship is between a moment or torque (instead of a force), angular acceleration (instead of linear acceleration), and mass moment of inertia (instead of mass). Therefore, similar to the linear case with $\vec{F} = m\vec{a}$, we can write an equation that describes the angular version as:

$$\vec{T} = I\vec{\alpha}, \tag{5.16}$$

where $\vec{T}$ is the torque, I is the mass moment of inertia, and $\vec{\alpha}$ is the angular acceleration vector. This means that the torque induces an angular acceleration in a body proportional to the mass moment of inertia that causes it to rotate. A larger torque creates a larger angular acceleration. However, if the mass moment of inertia is smaller, the angular acceleration will be larger for the same torque, and vice versa.

For example, consider a fan with the blades attached or removed. If the blades are not attached to the motor when it is turned on, the motor shaft rotates quickly in a very short period of time. This means that its angular acceleration is very high, and therefore it speeds up from zero to its maximum velocity very quickly. However, when the blade is attached to the motor and it is turned on it takes a relatively long time before the blades reach their maximum speed. Although it is true that the blades are also trying to push out the air and therefore add to the load on the motor, the much-lower angular acceleration is due to the much larger mass moment of inertia of the blades. Assuming that the torque of the motor is essentially the same, because the mass moment of inertia is larger, the angular acceleration is lower, requiring much more time to speed up to its maximum.

It may appear that the lower angular acceleration might just be the result of adding mass to the motor. However, if we were to add a metal ball with the same total mass equal to the blades to the motor and repeat the test, we would find that although the angular acceleration would be lower than no mass, it would still be much higher than with the blades. This indicates that it is *not* only the mass that matters but how it is distributed. In fact, we might mention that the rotor of the motor also has a mass moment of inertia that affects the angular acceleration of the rotor. Even when there are no blades attached to the rotor, the moment of inertia of the rotor is still present. We simply add to it when we attach the blades. The actual mass of the rotor may be much larger than the mass of the (plastic) blades, but the mass moment of inertia is much less compared to the blades. And this is why if we just add a ball with a mass equal to the blades, it will be as if the rotor were a bit heavier with little effect. But with the blades attached, the mass moment of inertia is increased significantly, affecting the angular acceleration significantly. So let's see how this can be analyzed and calculated. This analysis will help us understand what mass moment of inertia really is.

As shown in Figure 5.22, the mass moment of inertia for a plate of radius r, thickness t, and mass m about an axis x going through its center is:

$$I_x = \frac{1}{2}mr^2. \qquad (5.17)$$

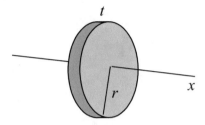

Figure 5.22: Mass moment of inertia for a plate.

Assume that a plate is 4 inches wide (2 inches in radius) and 1 inch thick. Its mass can be calculated by multiplying the volume by its density. The volume of the plate is its area multiplied by its thickness, or:

$$Vol = \pi r^2 t = \pi(2)^2(1) = 12.57 \text{ in}^3.$$

The specific weight of steel is 490 lb/ft^3 (0.2837 lb/in^3). This means that the density of steel is $\rho = 0.000734$ lbs^2/in^4 (this strange looking unit is the result of expressing the mass in English units). This is equivalent of $\rho = 0.00783$ kg/cm^3 in SI units. Therefore, the mass of the plate is:

$$m = Vol \times \rho = (12.57)(0.000734) = 0.0092 \text{ lbs}^2/\text{in}.$$

The mass moment of inertia of the plate is:

$$I_x = \frac{1}{2}mr^2 = \frac{1}{2}(0.0092)(2^2) = 0.0184 \text{ lbs}^2\text{in}.$$

Now let's take the same amount of material as before (same thickness, area, volume, and mass), but make it into a ring with an outside diameter of 5 inches and an inner diameter of 3 inches (outside and inside radii of 2.5 and 1.5 inches), as shown in Figure 5.23a. The area of the ring with r_o as its outer radius and r_i as its inner radius is:

$$Vol = t\left(\pi r_o^2 - \pi r_i^2\right) = 1\left(\pi(2.5^2 - 1.5^2)\right) = 12.57 \text{ in}^3,$$

which is the same as before, as will be its mass (0.0092 lbs^2/in). However, the mass moment of inertia of the ring is:

$$I_x = \frac{1}{2}m\left(r_o^2 + r_i^2\right). \qquad (5.18)$$

Substituting the new radii in this equation gives us:

$$I_x = \frac{1}{2}(0.0092)(2.5^2 + 1.5^2) = 0.0391 \text{ lbs}^2\text{in},$$

which is more than twice as large as the solid plate. Although we did not use any more material, simply by changing the size (a different distribution of mass), we more than doubled the mass moment of inertia.

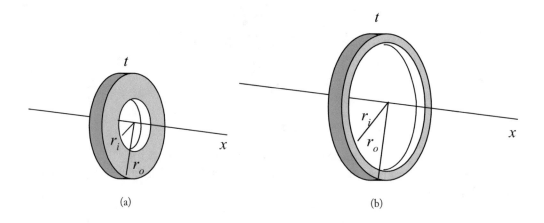

(a)   (b)

Figure 5.23: The mass moment of inertia of a ring is changed as the distribution of the material changes.

Now consider a third version: Assume we still use the same amount of material, but this time form the ring to have dimensions of $r_o = 5$ and $r_i = 4.58$ inches as shown in Figure 5.23b. In this case, too, since the area of the ring and its thickness are the same, so is its mass of $0.0092 \text{ lbs}^2/\text{in}$. However, the new mass moment of inertia will be:

$$I_x = \frac{1}{2}m\left(r_o^2 + r_i^2\right)$$
$$= \frac{1}{2}(0.0092)(5^2 + 4.58^2) = 0.211 \text{ lbs}^2\text{in},$$

which once again is nearly $0.211/0.0184 = 11.5$ times as large. This is the power of the way the mass is distributed. As the material is pushed outwardly, the mass moment of inertia increases. Figure 5.24 shows a typical way this is used in the design of machinery. In this figure, instead of attaching a uniform-thickness plate to the air motor, the same amount of material is made into the shape of a flywheel with a much larger mass moment of inertia. The larger moment of inertia is needed for smooth operation of the air motor, but it is provided without using a massive plate. In a typical flywheel, most of the mass is moved into the rim, which is connected to the hub with a thin plate. The same design is used in reciprocating internal combustion engines (used in all cars) to smooth out the variations in the thermodynamic cycle. See Chapter 4 for more discussion.

Figure 5.24: A typical flywheel is designed to have a larger mass moment of inertia without being too heavy by pushing most of the material outwardly to the rim.

If you were to turn on a fan with the blades attached, as in Figure 5.25a, you would notice that the blades take a while to reach their maximum rotational speed. However, if the blades were removed as in Figure 5.25b, the motor shaft would reach its maximum speed much more quickly, in a fraction of the time needed with the blades. As we discussed earlier, Equation (5.16) shows the relationship between the mass moment of inertia and angular acceleration. A fan motor without the blades has much less moment of inertia (of the rotor) than with the blades, especially since the mass moment of inertia of the blades, with their outwardly distribution, is relatively large. Therefore, the angular acceleration at the shaft without the blades is much larger than with the blades, and consequently, the motor reaches its maximum rotational speed much more quickly. This is even more apparent in ceiling fans, where the mass moment of inertia of the long blades is even higher.

Many bicycle enthusiasts look for a lightweight bike, usually at much higher cost, thinking that it is easier and faster to ride. Although the weight of the bike is a factor, the acceleration and how quickly the maximum speed is achieved are more importantly affected by the weight, size, and weight distribution of the tires. As you might guess by now, since bike tires rotate, their mass moment of inertia directly affects the angular acceleration, and consequently, how quickly the maximum speed is achieved. Therefore, skinnier tires used in racing bicycles that are lightweight

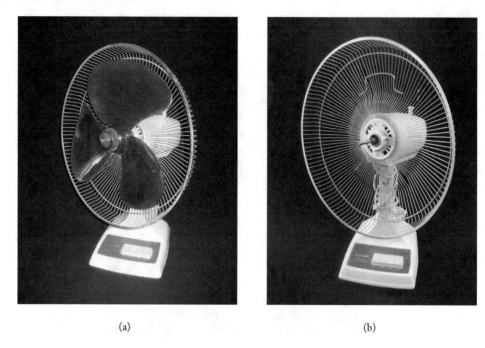

(a)   (b)

Figure 5.25: A fan motor accelerates much more slowly when the fan blades are attached as compared with the blades removed.

will have lower mass moment of inertia compared with fatter and heavier tires used in mountain bikes. Some bike owners go as far as drilling holes in the sprockets of their tires, thinking they are reducing the moments of inertia (as well as mass). How much effect do you believe this will have on the overall mass moment of inertia of the tires? Almost none. However, reducing the weight of the rim and the weight of the rubber used in the tire will significantly affect the moment of inertia.

Now let's look at a different situation. The propellers of airplanes and helicopters also rotate about the shaft, and like the aforementioned examples, we should expect their mass moments of inertia to affect the torque needed to rotate the propeller and the accelerations achieved. So first let's look at how we can calculate their approximate mass moments of inertia.

To do this, let's model the shape of a propeller as a slender bar. This approximation is useful for seeing what is important, but not accurate enough in practical terms. The actual mass moment of inertia can be found either experimentally or by writing more complicated equations.

Figure 5.26a shows a slender rod (the length is much larger than the diameter). Assume that the rod is attached to an axis at its center and rotates in a plane. Equation (5.16) still applies here; the applied torque is equal to the mass moment of inertia times the angular acceleration.

The mass moment of inertia of a slender bar is:

$$I = \frac{1}{12}mL^2, \tag{5.19}$$

where m is the mass and L is the total length of the slender bar. Now suppose that the bar rotates about one end, not the center. In this case, we need to calculate the moment of inertia about the end, not about the center. As with the second moment of the area about an axis other than the centroidal axis, we need to use the parallel axis theorem to calculate the mass moment of inertia about an axis other than the one at the center of the mass. This, similar to Equation (5.6) for area moment of inertia, can be written as:

$$I_B = I_o + md^2, \tag{5.20}$$

where I_B is the mass moment of inertia about an axis B away from the center of mass, I_o is the mass moment of inertia about the center of mass, and d is the distance between the two axes. Therefore, for the slender bar of Figure 5.26b, the mass moment of inertia about one end will be:

$$I_B = I_o + m\left(\frac{L}{2}\right)^2$$
$$= \frac{1}{12}mL^2 + \frac{1}{4}mL^2 = \frac{1}{3}mL^2.$$

Obviously, this moment of inertia is 4 times as large, resulting in an acceleration that is 4 times as slow, or requiring 4 times as large a torque to rotate the bar at the same rate.

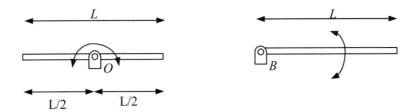

Figure 5.26: The mass moment of inertia of a slender bar.

Just to clarify this in a different way, let's recalculate the mass moment of inertia of the slender bar about its center of mass by assuming that it is the summation of two bars, each with a length half as much and a mass half as much, attached together at one end. Therefore, the total mass moment of inertia will be twice the moment of inertia of a bar at half the length and half the mass, calculated at its end, or:

$$I_O = 2\left(I'\right) = 2\left(\frac{1}{3}\left(\frac{1}{2}m\right)\left(\frac{L}{2}\right)^2\right)$$

$$= \frac{1}{12}mL^2,$$

which is exactly the same as before for the mass moment of inertia of a slender bar about its center of mass.

Therefore, when propellers are longer or heavier, their mass moments of inertia increase. In helicopters, where the propellers are much longer than in airplane engines, it is almost impossible to turn them as fast as in an airplane; their mass moment of inertia is much larger, putting a much larger load on the engine.

As you can see, both the second area and mass moments of inertia play a fundamental role in many things in our daily lives. For example, you can see the effects of the moment of inertia of the wings of a bird both in terms of their strength under the weight of the bird as well as how much force (or moment) is needed to flap them and in the effects of the moments of inertia of the legs of different creatures, including humans, in running. You can hopefully imagine these same effects considered in the design of a bridge, the flight of an airplane, the rotating parts of an engine, and countless other devices and machines we use every day. Understanding these concepts helps us both control their effects and use them to our advantage.

CHAPTER 6

Electromotive Force

Motors, Transformers, AC and DC Currents

6.1   INTRODUCTION

Each of the two generators of the Diablo Canyon nuclear power plant in San Luis Obispo County generates over 1,100 megawatts of power, enough for about 3 million people. The Ames Research Center national full-scale subsonic wind tunnel in Mountain View, California, is 40 × 80 ft and creates winds of up to 350 mph (560 km/h), large enough to test a real, full-scale Boeing 737. The fans and the motors running these fans are enormous. And yet, the generators used to recharge a hand-held flashlight are the size of a large olive and the motors used in small remote-control servomotors are about ¼ inch in diameter. What is important is that the largest and the smallest of motors and generators are actually very similar in the way they work and that they all follow Faraday's Law which we will study later.

When you simply plug in an electric motor (whether as part of a device or stand-alone) it simply turns and provides a torque that allows the device to do its job. The same is true for a DC motor that is connected to a battery. You may also use a simple charger (or transformer) to recharge your batteries, whether in a cell phone, camera, computer, hybrid car, or toy. In fact, you may have heard that the high voltage (as large as 500,000 volts) of electric power is lowered to the household voltage (110 volts) with a transformer before it is delivered to your place of residence or work. All these examples are based on a phenomenon called *electromotive force* or *emf*. A similar phenomenon that works in the opposite way, called *back-emf*, is also an important issue that affects the way these systems work or are designed.

In this chapter we will study these two phenomena, how they are used, and where they appear to affect our daily lives. But first let's learn the difference between voltage and a current, and their relationship. These terms appear in all issues related to circuits and electric devices.

6.2   INTRODUCTORY TERMS: VOLTAGE, CURRENT, AND RESISTANCE

Equation (6.1) shows the relationship between voltage (V), current (I), and resistance (R). But what is the physical meaning of these terms? To understand it, let's make an analogy. We will look at a simple fluid system to show how they are related.

$$V = IR. \tag{6.1}$$

Imagine a tank of water with a pipe attached to it, full of water as shown in Figure 6.1. At the bottom of the pipe there is a valve, closed at this time, which prevents the water from flowing in the pipe. The pressure at the bottom of the pipe is a function of the density of water and its height. Larger heights (h) will increase the pressure at the bottom of the pipe.

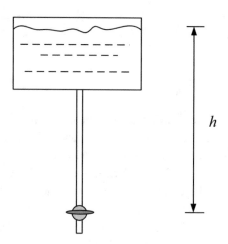

Figure 6.1: A tank-pipe-valve system shows the analogy between hydraulic and electric systems.

Now imagine that we open the valve just a bit. As a result, water will start to flow slowly at the bottom. The amount of water flowing is a function of the pressure and the opening of the valve. At this point, the valve provides resistance to the full flow of the water. Further opening the valve will increase water flow until it is fully opened, at which time the flow is at its maximum rate. Obviously, at higher pressures, the flow will be larger too. Notice that regardless of the valve opening, the flow is also a function of the diameter of the pipe. Smaller diameters provide more resistance to the flow. For example, if the pipe were a hose with a small diameter, the flow would be less than if it were a large pipe. Therefore, the pipe diameter also introduces resistance to the flow.

Electrical systems can be explained the same way. The water pressure is analogous to the voltage. The flow of the water is analogous to the current which represents how many electrons pass a cross section of the wire. The valve represents a variable resistance similar to resistors that are used in circuits to control current. The resistance of the pipe too represents the electrical resistance of wires and conductors. As shown in Equation (6.1), when the resistance of a circuit increases, the current decreases. By changing either the voltage or the resistance, the current flow can be controlled. For a system (for example a motor) where the resistance is constant, when the voltage changes, the current changes accordingly. In mechanical systems the equivalent analogous elements are force and voltage, velocity and current, and viscous coefficient of friction and resistance. To understand this, think of walking in a pool. You need to exert yourself to move in the water. The thicker the fluid, the harder it will be to move.

6.3 MAGNETIC FIELDS

Imagine that there is a magnetic field present. This can happen if you have a permanent magnet or if you wrap a wire into a coil and run a current through it, as shown in Figure 6.2. In the vicinity of the coil or the magnet, there will be a magnetic field such that if we bring another magnet close to either of them, similar poles (both north or both south) repel each other and the opposite poles attract even before they touch. Since the strength of the field (flux density) at any point, among other things, is related to the square of the distance from the source, the field strength reduces quickly as you move away from the source. Therefore, with most simple magnets, this is felt when you are close. The same can be felt at larger distances when you experiment with a stronger magnet.

Figure 6.2: A permanent magnet and coils (when electricity flows through them) create a magnetic field around themselves.

You can actually visualize a magnetic field by peppering small pieces of iron (filings) onto a piece of paper over a magnet as shown in Figure 6.3a. The lines formed by the iron filings show the shape of the field between the poles. Figure 6.3b shows the general shape of magnetic fields.

Magnetic flux lines always close between the poles. Unless they are somehow concentrated by other means, they surround the magnet in all directions. This is similar to having a source of light that illuminates in all directions equally. As a result, the strength of the magnetic field is distributed and low. However, like a flood light whose light intensity is concentrated in a small angle by reflectors, the strength of the magnetic field around a magnet or a coil can be increased locally by concentrating them with an iron core. This is why we almost always see an iron bar within a coil; the coil generates the electromagnetic field, but it is distributed all around and consequently, it is very weak. The iron core inside it concentrates the field causing it to be much

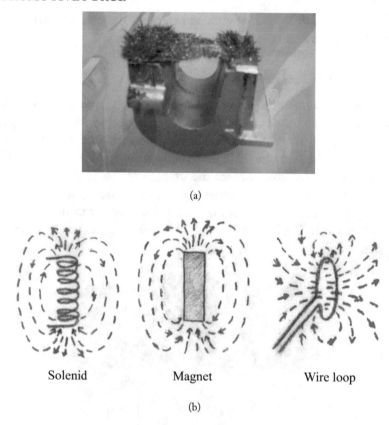

(a)

Solenid   Magnet   Wire loop

(b)

Figure 6.3: The shape of a magnetic field around magnets and coils. Iron filings, peppered within the field will line themselves up to follow the magnetic flux lines.

stronger around the bar. For the same reason too, all motors are invariably made with a metal casing to concentrate the magnetic field within the casing. As a result, even if you bring a small piece of steel near the body of a motor, the magnet within the motor does not attract it. This is an important characteristic of magnets and coils and is used in almost all transformers and motors as well as magnetic devices used as sensors such as a *Linear Variable Differential Transducer* (*LVDT*) used to measure distances. We will later discuss the additional effects of the iron cores in a transformer, how they can be a detriment to the efficiency of the system, and how we can overcome that.

Before we explore the subject of electromotive force, let's first study magnetic flux a bit more. This will help us understand the subject much more easily.

The strength of the magnetic flux is the product of the flux density (the level of the concentration of the magnetic flux in any area) and the area, or:

$$F = B \cdot A, \tag{6.2}$$

where F is the flux, B is the flux density, and A is the area. As expected, the flux changes as a result of any variations in either the flux density or the area. As we will shortly see, what is important in the generation of electromotive force is not just the strength of the field but the changes in it with respect to time.

These changes come about as a result of the changes in the flux strength or the area. For example, in a transformer the strength of the flux (flux density) changes due to the nature of the alternating current (AC) power. The AC current or voltage follows a sine function as shown in Figure 6.4. The voltage changes from zero to a maximum level, then decreases back to zero, then follows to a maximum with the opposite polarity (negative direction), finally returning to zero again, and repeating the pattern 50 times a second (or 60 times a second outside of the U.S. and Canada). As the voltage changes, so does the flux density. As a result, when a coil is connected to AC power, its flux density varies continuously.

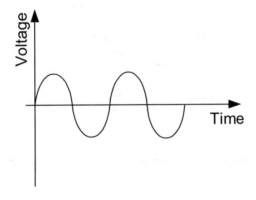

Figure 6.4: The sinusoidal nature of AC power creates a continuous change in the flux density of a coil.

Now imagine that we have a permanent magnet or a coil (connected to a DC source of power which does not vary) with constant flux density. If we pass a conductor (e.g., a piece of wire) through the flux, the conductor will disrupt the flux, effectively changing its area. This means that as a result of passing the conductor within the flux and crossing its lines, we cause a change in its area, creating variations in the flux. The effect is the same as when the flux density changes. Either one of these two will change the flux strength.

6.4   ELECTROMOTIVE FORCE

Electromotive force relates to the interactions between a magnetic field and an electrical current through a conductor (such as a wire). According to *Faraday's Law,* when there is a change in the magnetic field, the interaction results in the generation of a force if the conductor is carrying a current, or induction of a voltage (or current) in a closed-loop conductor if it is moved (by a force). A series of simple experiments by Michael Faraday in England and by Joseph Henry in the U.S. in 1831 helped formulate this phenomenon. Figure 6.5 shows a galvanometer (which measures a current) in series with a simple conductor coil. If a bar magnet is moved toward the coil, the galvanometer deflects, indicating a current in the conductor. If the bar is stopped, the galvanometer goes back to zero. If the bar moves back, the galvanometer deflects in the opposite direction, indicating a current in the opposite direction. If the magnet is reversed, all these indications also reverse. The same will happen if the magnet is kept stationary but the coil is moved relative to it. Therefore, as shown, when the magnet is moved within a coil it generates a current. This is called an induced *electromotive force* or *emf.*

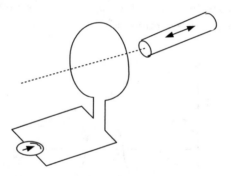

Figure 6.5: A magnet moving toward a coil induces a current in the coil. This is called induced *electromotive force* or *emf.*

Similarly, as shown in Figure 6.6, if the galvanometer and the coil are stationed close to another stationary coil that is connected to a power source (battery), when the switch is closed or opened the galvanometer deflects momentarily in opposite directions, but not if the switch is left on or off. It is only as a result of the switch turning on or off that the galvanometer deflects, indicating that the change in the current causes an electromotive induction in the coil. This is the principle behind the generation of electrical power in a generator.

This interaction between a magnetic field, a conductor, and relative motion (caused by a force that creates the motion) is interchangeable. This means that as in Figure 6.5, in the presence of a current through the coil, the magnet will experience a force that moves it relative to the coil (still called the electromotive force). The same principle is the basis on which all motors operate too.

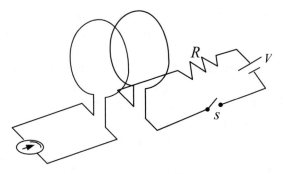

Figure 6.6: Whenever the switch is closed or opened the galvanometer deflects, indicating a momentary induction of electromotive force in the coil.

The opposite of the same phenomenon is called *back electromotive force* or *back-emf*. We will discuss this a bit more later. Now let's see what this means in practice.

Notice that as we just saw in Section 6.3, the change in the flux can come from a change in its density or from a conductor crossing its lines. Also note that a closed-loop conductor means that the wire is continuous or attached to a load. For example, let's say we attach the two sides of the wire to a lamp. In that case, the circuit is closed-loop or continuous, and therefore, the voltage which is induced can travel through the wire to the load (lamp) and return. This creates a current in the wire. Otherwise, if there is a voltage but the wire is not continuous or attached to a load, there is no current flow and nothing happens.

> As a side note, I remember a group of first-year students who had designed and constructed a device which, based on this principle, was supposed to reduce vibrations in a pendulum as it passed through a magnetic field. However, the device was not working and the students had assumed that it was not constructed well. What they had not realized was that since the pendulum was insulated and not attached to a load to actually use the voltage induced in it, there was no current and as a result there was no damping of the pendulum.

As mentioned previously, this is the principle that governs the operation of all motors, generators, and transformers. Although these are seemingly different devices, the different interactions of the same elements of Faraday's Law are at work for each one. Now let's see how each one works.

First let's see about motors. Imagine that there is a magnetic field generated by a permanent magnet (where the flux intensity is constant). Now imagine that we take a conductor such as a wire and pass a current through it. As a result of Faraday's Law, the interaction between the flux and the current-carrying conductor is a force on the conductor, pushing it away. We will see how an actual motor works continuously, but for now, as you can see, a force is generated that pushes away the conductor.

Now take the same system mentioned previously, but instead of supplying a current through the conductor, assume that we move the conductor through the flux (which is caused by a force we supply). The crossing of the flux also creates a change in the flux (area), and based on Faraday's Law, this will induce a current in the conductor. This system is a generator (and we will see the details later). Note that a generator and a motor are the same; in one, we supply the current and it moves, in the other we supply the motion and it induces a voltage (or current). It should be mentioned that although the workings of a DC and AC motor and their details are different, they all follow the same principles.

Next consider a transformer. There are no moving parts in it. Instead, two coils interact with each other. The supplied current to one coil is AC power which varies constantly and changes the flux. Consequently, as a result of the changes in the flux and based on Faraday's Law, a voltage is induced in the second coil. Therefore, although these devices are different and each one is designed for a different application, they all follow the same rules. Now let's see how AC and DC motors, generators, and transformers work.

6.5 DC MOTORS

DC stands for Direct Current, meaning that the polarity (direction of flow) does not change. If a battery is used, the magnitude does not change either. DC power is usually supplied by batteries or by circuits that deliver a direct current. Therefore, a DC motor requires a DC source and will not work with AC power. These motors are powerful, their direction of rotation can be changed, and their speed can be controlled relatively easily as we will discuss later. However, their power-to-weight ratio is lower than AC-type motors and they cannot tolerate high temperatures as much as AC-type motors.

As expected, DC motors operate based on the principles of electromotive force. A permanent magnet (called a *stator*) provides a magnetic field whose lines are crossed by a current-carrying conductor (called a *rotor*). Since the effective area of the flux is disrupted, a force is generated that pushes the coil rotor as shown in Figure 6.7a. Since the conductor is attached to a shaft, it rotates to the middle as a result of the force. The same thing happens to the second coil which is now receiving the current. Therefore, the rotation continues. In order to provide the current to the coils sequentially when needed, a set of commutators and brushes are used. The coils are connected to the commutators. The brushes, carrying the current from the power source, slip over the commutators and supply the coils with a current.

In reality, the rotor coils are formed around iron cores, creating magnets. We may describe the motion of a DC motor through the pulling of opposite poles and pushing of similar poles between the poles of the stator magnet and the rotor magnets. When the current is supplied to a coil, the core develops a north pole and a south pole that are pulled or pushed by the poles of

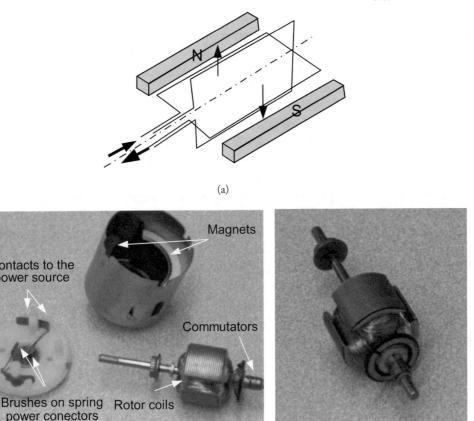

(a)

(b)   (c)

Figure 6.7: The components of a DC motor.

the stator, causing the rotor to rotate. However, as soon as one coil rotates away, power is cut off and instead, the next coil is powered which repeats the process until the power is cut off. In order to make the motion of the motor smooth, most DC motors have rotors with three coils (Figure 6.7c). Either one or two of the three coils are powered at any given time. Figure 6.7b shows the rotor, stator, commutators, and brushes of a small DC motor.

6.6   AC MOTORS

AC motors are simpler in their construction and operation and are therefore more rugged. They are made of a permanent magnet (PM) rotor or a simple cage (such as *squirrel cage rotor*) and a coil stator. AC motors do not have brushes or commutators because AC power automatically provides a changing magnetic field and consequently, there is no need for external switching of the direction of the current for continuous rotation.

Imagine that a permanent magnet is mounted on a shaft to form a rotor as shown in Figure 6.8. The stator is a coil in which the current flows. Let's assume that the rotor is situated such that both the north and south poles are aligned with poles of the coil which at this instant is not magnetized yet. Imagine that at the instant shown, the AC current starts from zero in the positive direction flowing into the coil. This will create a magnetic field such that both north poles and both south poles of the rotor and stator will repel each other, forcing the rotor to rotate in the direction shown in order to bring the north of the rotor closer to the south of the stator and the south of the rotor to the north of the stator. As the AC current increases to its maximum level, the repelling and attracting forces between the poles increase. Eventually, the AC current starts to decrease toward zero, but still provides the same forces in the same directions. As soon as the poles of the rotor and stator align themselves with each other's opposite poles, the direction of the AC current changes its polarity, thereby switching the directions of the poles on the stator. This will create the exact situation as before, but at the new location, repelling the now-similar south poles and north poles toward the opposite side. With this complete cycle, the rotor rotates once. But as soon as it reaches the opposite side, the AC switches again, forcing the motion to continue, repeating indefinitely until the motor is shut down. There is no need for commutation, switching, or brushes.

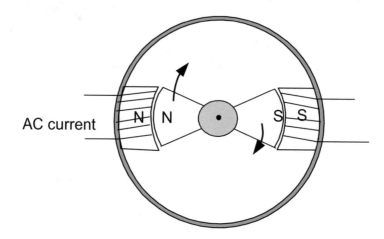

Figure 6.8: A permanent magnet AC motor.

Notice how the rotor follows the stator's moving field. Due to the nature of AC power, the field continually switches directions, and the rotor follows it. Therefore, the speed at which it rotates is a function of the line frequency. For example, the line frequency in the U.S. is 60 Hz. Depending on the number of poles used, the speed of an AC motor with permanent magnets will be 1,800, 3,600, etc. The same motor in many other countries whose line frequency is 50 Hz will be 1,500, 3,000, etc. These are usually referred to as *synchronous motors* because they have a fixed speed that is a function of the line frequency. This speed, to a large extent, stays constant as the load increases or decreases, but the angle between the rotor and stator changes a little to increase or decrease the load. If the load increases more than the motor can handle, instead of slowing down as a DC motor does, it simply stops.

Another type of AC motor is called an *induction motor*. Induction motors are very similar to synchronous AC motors but instead of the permanent magnet rotor, they simply have a metal rotor in the form of a number of longitudinal bars that are connected together like a cage. Therefore, the rotor is usually called a *squirrel cage rotor* as shown in Figure 6.9. Notice that the rotor is not powered by any electrical current; it is simply a metal cage. Similar to other AC motors, the stator is made of coils that are powered by an AC current. In this case, due to back-emf, the varying flux induces a current into the cage. However, since there is a current in the cage's conductors, a force is generated that rotates the rotor. However, in this case there are no poles that exactly follow the magnetic field, and consequently, the rotor can rotate at any speed as the torque and current change. These motors, also called *asynchronous motors*, are rugged, powerful, long lasting, simple, and economical. There are no brushes, magnets, or commutators. They are used in most AC applications. Their basic disadvantage is that they always rotate in the same direction. Therefore, they can only be used in applications where the motor does not need to change direction (e.g., washing machines, dryers, fans, pumps, etc.). They cannot be used as drill motors because unlike DC motors, they cannot be reversed.

Figure 6.9: An AC induction motor's stator and rotor. The rotor is a simple non-magnetic cage with no power supplied to it. AC power is supplied to the stator coil only.

One exception is called a *reversible AC motor*, where the coil is center-tapped as shown in Figure 6.10. In this case, only 1/2 of the coil is used for each direction, producing 1/2 of the torque. Therefore, for the same power rating, these motors need twice as much winding, making them heavier and more costly. To reverse the direction of rotation, the AC current flows from the center to only one side of the coil, thereby going left to right or right to left, creating a field that is the opposite of the other case. As a result, the rotor will rotate in one direction or another.

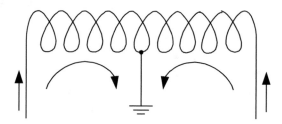

Figure 6.10: A reversible AC motor can switch directions because the stator coil is center-tapped. As a result, the current flows in opposite directions depending on which route is chosen, creating fields that are in opposite directions and forcing the rotor to rotate in opposite directions.

For devices like drill motors that require direction change but where DC power is not available unless it is rectified and lowered to suit common low-voltage DC motors, a *universal motor* is used. Universal motors are a combination of DC principles and AC power. Instead of permanent magnet stators like those in DC motors, they have coil magnets as in AC motors that need to be supplied with an AC current. The rotor is a coil with brushes and commutators that is also powered by an AC current. In this case, the magnetic field caused by AC power in the stator coil changes direction 60 times a second, but because the rotor is also supplied with the same AC current, its direction also changes the same 60 times per second at precisely the same time. Since they both switch directions at the same time, it is as if it were a DC current. The additional switching through the brushes and commutators causes the rotor to rotate like a DC motor. Therefore, although powered by an AC current, the motor's direction of rotation can be switched like a DC motor.

In other types of motors such as stepper motors and brushless DC motors, the attempt is to do the opposite; to run a DC motor with the construction of an AC motor with no brushes or commutators. This makes the motors more rugged and longer lasting. We will study these motors next.

6.7   STEPPER MOTORS

Unlike DC and AC motors that start rotating continually when they are connected to a power source, stepper motors do not; they only move one step when the field This field rotation is usually accomplished by a dedicated circuit, a computer, a microprocessor, a PLC (Programmable Logic Controller) or similar means. Therefore, the movements of the rotor are under complete control of an external device.

To understand the way a stepper motor works let's consider a simplified version first. Figure 6.11a shows a permanent magnet rotor and a coil in off position with their poles aligned. As soon as the coil is turned on, the similar north and south poles will repel each other (Figure 6.11b) until the poles of the magnet line up with the opposite poles of the coil (Figure 6.11c). At this point, the rotor will stay here without movement, even if we turn off the coil. This is called the point of *least reluctance*, a stable position. In this position, even if we apply a torque to move the rotor it will resist. If we once again turn on the coil in the opposite polarity of the first case, the similar poles will repel each other again (Figure 6.11d), forcing the rotor to rotate again until the opposite poles align and it stops again. In this process, every time we turn on the coil in proper polarity, we force the rotor to rotate half of a full circle or 180°. However, although we can force the rotor to only rotate this fixed amount, there are two problems with this set up: (1) that the step size is large, and (2) there is no control over the direction of motion that the rotor takes. When we turn on the coil, the rotor may rotate either clockwise (CW) or counter-clockwise (CCW).

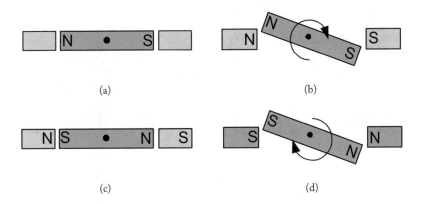

Figure 6.11: A simple stepper motor set up.

To improve this situation let us consider the set up in Figure 6.12 where we have added a second coil. In this case, assume that we start as before, when both coils are off and the rotor is aligned with the poles of coil-1. Now assume we turn on coil-2 such that its poles will be as shown in Figure 6.12b. As a result, the rotor will rotate to align itself with the poles of coil-2. If we next turn off coil-2 and turn on coil-1 as shown in Figure 6.12c, the rotor will rotate until its poles align with the poles of coil-1. Once again, we turn off coil-1 and turn on coil-2 in the opposite polarity, forcing the rotor to rotate again. Continuing to turn on and off coils-1 and -2 sequentially in proper polarity we can force the rotor to rotate clockwise or counterclockwise as much as we want. In this case, the step size is reduced to a quarter of a turn or 90° and we know the rotor's direction of rotation; unlike the first case it is not left up to chance, which is a big improvement. Also notice that by selecting how many times we turn each coil on and off we can ensure that the stepper motor rotates exactly as much as we wish, no more, no less. Additionally, by selecting how fast we turn the coils on and off we can control how fast the rotor rotates; if they are turned off and on more quickly, the rotor will also rotate more quickly and vice versa. So we are in complete control of the magnitude of rotation, the speed of rotation, as well as the direction of rotation.

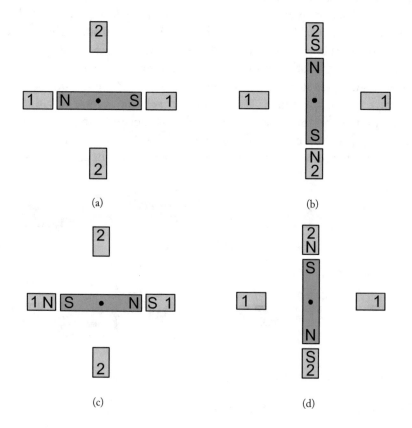

(a)   (b)

(c)   (d)

Figure 6.12: Employing two coils instead of one will improve the stepper motor behavior and characteristics.

We can improve the situation and cut the size of the step in half if we employ another variation. As shown in Figure 6.13b, suppose that instead of turning off coil-2 at this instant and turning on coil-1 we would keep it on while we turn on coil-1. With both coils on, since the rotor's magnet needs to balance itself at the point of least reluctance, it will align itself in the middle of the arc between the two, thereby rotating only 45° (Figure 6.13c). If we then turn off coil-2 it will rotate the remaining 45° to complete the step (Figure 6.13d). This is called *half stepping* and is common in many applications. Therefore, without adding any new coils or other components, simply by controlling when the coils are turned on and off we can reduce the step size by half. The only remaining problem is that for most applications, even 45° is a large displacement. This is because although we have control over displacement, speed, and direction of rotation, we actually have no control over the location of the rotor in between the poles when it is under load. Therefore, it is desirable to reduce the size of these steps even further. However, there is a limit to how many coils we can add. In the following section, we will see how two different methods are employed to reduce the step size of common stepper motors significantly without adding a significant number of coils.

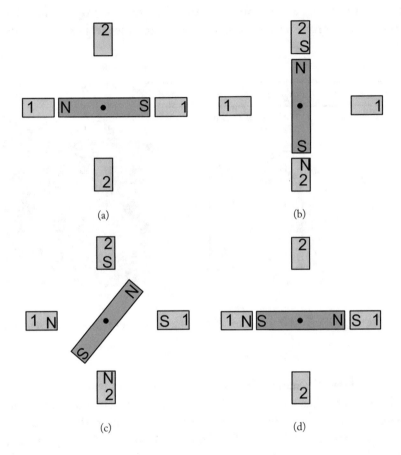

Figure 6.13: Schematic of an improved stepper motor.

6.7.1   CANSTACK STEPPER MOTORS

Canstack motors are rugged and simple in construction. The motor is comprised of a permanent magnet rotor made of a flexible sheet magnet that is similar to the type that is used for refrigerator magnets. These magnets, called *halfback array magnets*, are made by embedding (powder) steel filings in a flexible resin and magnetizing them with a machine. To understand the difference between these magnets and a regular steel magnet, turn one of these magnets on its back and try to stick it to any steel material. You will notice that the magnet does *not* stick. This is not due to the fact that many of these magnets have a sheath of plastic, used for advertising, on them. It is because these magnets are magnetized only on one side. Figure 6.14 schematically shows how these magnets are a series of magnets next to each other with all their poles on one side; the opposite side is not magnetic. Figure 6.14 also shows the construction of the rotor of the stepper motor and a real rotor. The rotor is made of the same type of magnet, rolled into a cylinder. Therefore, the rotor will have a series of south and north poles sequentially located next to each other.

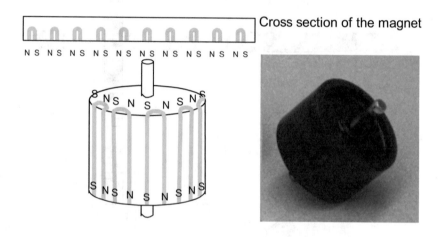

Figure 6.14: The magnetic field of the rotor of a canstack stepper motor possesses a series of magnets next to each other.

Figure 6.15 shows the stator of a canstack stepper motor. It is made up of two electromagnets, stacked over each other, each made up of two plates and a coil. Each plate has, in this case, 12 little fingers or tabs on it as shown. When a current flows in the coil, each of these plates becomes either a north or douth pole. Therefore, there will be 12 tabs of north and 12 tabs of south created when a current flows in each coil. The coils can be turned on and off independently of each other in either polarity by center-tapping the coil as was discussed in Section 6.6, Figure 6.10. Notice that this means that when a coil is turned on, it creates an equivalent of 12 magnets (24 poles), or a total of 48 tabs and 48 poles around the stator when both coils are turned on. But instead of having to make 24 individual coils within the motor and turning them on and off sequentially, we only need to turn two magnets on and off. But notice that although we only have two electromagnets, since each coil is center-tapped, we effectively have four coils, referred to as Coil-1, Coil-2, Coil-3, and Coil-4. Coil-1 and Coil-2 are the same coil, but with opposite polarity, etc.).

Figure 6.15: Canstack stepper motor is comprised of a permanent magnet rotor and a stacked, two-stage stator with repeating poles that are staggered from each other to provide small step sizes.

Let's call the plates (and thereby the tabs attached to each) *A* and *B* for the first electromagnet (Coil-1 and Coil-2) and *C* and *D* for the second electromagnet (Coil-3 and Coil-4). Table 6.1 shows their magnetic poles for each polarity:

Table 6.1: The poles of the stepper motor electromagnets versus the polarity of the current

	Tabs *A*	Tabs *B*	Tabs *C*	Tabs *D*
Coil-1	N	S		
Coil-2	S	N		
Coil-3			N	S
Coil-4			S	N

The trick is that these plates are assembled such that the tabs form a staggered set so that they will have a sequence of *A, C, B, D, A, C, B, D, A, C,*. . . . . Figure 6.16 shows this arrangement in a linear fashion.

So, what is the effect of this arrangement? Suppose that we turn on Coils 1 and 3 at the same time. This means tabs *A* and *B* will be N and S, and tabs *C* and *D* will be N and S. Therefore, the sequence of *A, C, B, D, A, C, B, D, A, C,*. . . will result in N, N, S, S, S, N, N, S, S, etc. (please follow this carefully). Similar sequences will form as we turn the coils on and off. Table 6.2 shows the pattern when the coils are switched six times.

Notice that in Table 6.2, as highlighted, the field rotates in one direction. For example, any two south poles next to each other advance one step as the coils are switched on and off

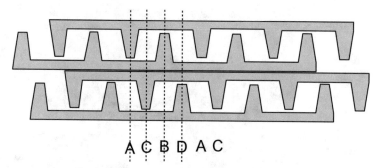

A C B D A C

Figure 6.16: The four plates that constitute the two magnets of a canstack stepper motor are staggered relative to each other such that their tabs are sequentially repeating in an *A, C, B, D, A, C, B, D, A, C,.......* fashion.

Table 6.2: The sequence of magnetic poles of stepper motor tabs as the coils are sequentially turned on and off

	A	*C*	*B*	*D*	*A*	*C*	*B*	*D*
Coil-1, Coil-3	N	N	S	S	N	N	S	S
Coil-1, Coil-4	N	S	S	N	N	S	S	N
Coil-2, Coil-4	S	S	N	N	S	S	N	N
Coil-2, Coil-3	S	N	N	S	S	N	N	S
Coil-1, Coil-3	N	N	S	S	N	N	S	S
Coil-1, Coil-4	N	S	S	N	N	S	S	N

sequentially, as do the rest. Therefore, if the rotor is aligned with the tabs such that its north is between the two souths, the rotor moves with the sequence as shown in Figure 6.17.

The continuous motion of the stepper is accomplished by simply repeating the sequence of switching coils 1 through 4 in 1-3, 1-4, 2-4, 2-3 order as shown in Table 6.2. Reversing the sequence will force the rotor to turn backward. Since there are 48 tabs, each step will be 360 /48 = 7.5 . Consequently, the total rotation of the rotor will be equal to 7.5 n where n is the number of switchings. The faster we switch, the faster the rotor rotates. This way, we are in complete control of the total angular displacement, angular speed, and direction of rotation of the rotor.

If a microprocessor is used to run the stepper motor, as is the case in most devices, the microprocessor turns four switches on and off that provide a current to each coil in a 1-3, 1-4, 2-4, 2-3 sequence, which is extremely simple to program with a microprocessor. Figure 6.18 shows a schematic of this set up.

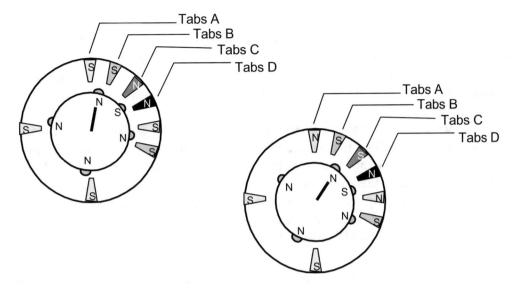

Figure 6.17: The rotor follows the field as the field is advanced in the stator of a stepper motor. For simplicity, only some of the tabs are shown.

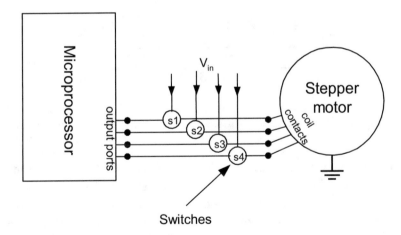

Figure 6.18: The schematic of a simple set up to run a stepper motor with a microprocessor.

It should be mentioned here that there is much more to stepper motor drives, control schemes, efficiency, and other issues that are beyond the scope of this discussion.

6.7.2   HYBRID STEPPER MOTORS

These stepper motors usually have a much smaller step size, for example 1.8° at full step and 0.9° at half step. This translates to 200 and 400 steps per revolution respectively. However, this is achieved with the same number of coils. To understand how this works, let's look at a simple principle that is not only used here, but also in calipers that are used for measuring dimensions more accurately.

Imagine that a bar A with a certain length is divided into 10 portions as shown in Figure 6.19. Obviously, each portion will be 1/10th. Also imagine that bar B with the same length is divided similarly into 10 portions. Therefore, the divisions of both bars A and B will be exactly the same. If at any time the division marks of A and B are aligned, one of the bars would have to move one full division in order to align the next set of division marks. For example, if division marks 2 on A and B are aligned, the next possibility for alignment will be if B is moved one full division until 3 on A will be aligned with 2 on B.

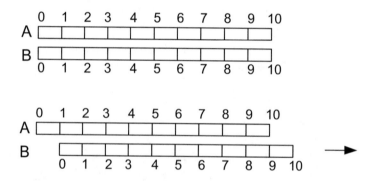

Figure 6.19: Aligning division marks of similar lengths requires one full-length motion.

Now imagine that we take the same length bars A and B, divide A into 10 portions as before, but divide bar B into 11 divisions as shown in Figure 6.20a (other numbers of divisions are perfectly fine too. Each number will result in a different proportion, but they are all fine). Also imagine that at one point, division mark 3 on bar A is aligned with division mark 3 on B (Figure 6.20b). Unlike the previous case where the division marks were all the same length, here they are slightly different. Consequently, all it takes to align the next set of division marks is for bar B to move only about 1/10th of this distance, or about 1/100th of the total length until division mark 4 on A aligns with division mark 4 on B (Figure 6.20c). In other words, since the divisions are no longer the same, bar B only has to move the distance of the difference between the two, or $(1/10) - (1/11)) \cong 0.01$. This means that we have made the divisions so much smaller without having to draw 100 lines. This principle is used in calipers to measure dimensions to about 1/100th of an inch. It should be mentioned here that it does not matter what the number of divisions are as long as they are not equal. So we could achieve similar results (albeit different values) with 10 and 8, 8 and 7, 20 and 21, or any other unequal pairs.

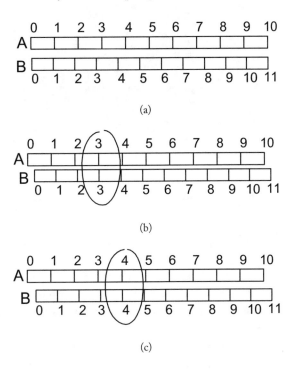

Figure 6.20: The unequal number of divisions on equal lengths allows for measuring much smaller distances as in a caliper.

Figure 6.21 shows the construction of the rotor and stator of a hybrid stepper motor. Notice how the rotor and the stator have teeth or divisions on them. In this particular example, the rotor has 50 divisions, and the stator has the equivalent of 40 divisions. Just like the caliper, in order to move the rotor to align with the next division at any location, the rotor has to only move an angle of $[(1/40) - (1/50)] \times 360° = (0.025 - 0.02) \times 360° = 1.8°$, which is much smaller than the canstack step size. Combining these two seemingly unrelated concepts benefits us very much.

The rotor of a hybrid stepper motor is a simple magnet with one north and one south pole. To reduce the back-emf current in it as it rotates, the rotor is made of laminated layers attached together to form the rotor as is shown in Figure 6.21. The stator has four coils in it that can be individually turned on and off. Therefore, exactly like the canstack motors and with a similar schematic as in Figure 6.18, a sequence of 1-3, 1-4, 2-4, 2-3 or its reverse drives the hybrid stepper motor forward or backward with complete control over its displacement, speed, and direction.

Figure 6.21: A hybrid stepper motor and its rotor and stator construction.

6.8   TRANSFORMERS

Transformers are used to increase or decrease voltages and currents. They function based on the same principles we have already discussed although there are no moving parts in them. They are composed of a primary coil, a secondary coil, and an iron core to concentrate the flux and increase the efficiency of the device.

The primary and secondary coils are simple coils with different number of turns (loops) in them designated as N_1 and N_2. An AC current is fed into the primary coil which creates a varying flux. According to Faraday's Law, since the flux intensity is changing due to the AC current, it induces a voltage into the secondary coil. In general, without an iron core to concentrate the flux, the level of induced voltage is very low and most of the energy is wasted. However, in the presence of a core, the efficiency of the system can be increased to 90% or better. Figure 6.22 shows the schematic drawing of two ways transformers may be built. Figure 6.23 is a typical transformer. The primary and secondary coils, the iron core, and the connections for different levels of voltage can clearly be seen.

The induced voltage can be expressed as:

$$V_{out} = V_{in}\alpha \left(N_2/N_1\right) \cos \beta, \tag{6.3}$$

where V_{out} is the voltage in the secondary and V_{in} is the supplied voltage in the primary. The constant α represents the effects of the coupling between the primary and secondary coils as a result of the iron core concentrating the flux and can vary from near zero to a maximum of 1 under the best conditions. Larger values of α indicate a better and more efficient transformer. β is the angle between the primary and secondary coils. N_1 and N_2 are the number of turns in the primary and secondary coils.

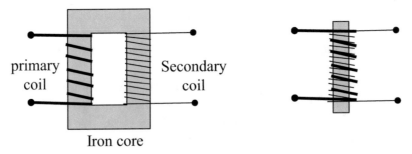

Figure 6.22: Schematic drawing of two ways a transformer may be built.

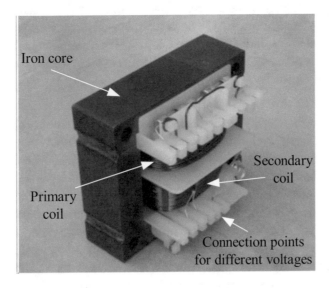

Figure 6.23: A typical transformer with its coils, iron core, and connections.

Varying the V_{in} will proportionally change V_{out}. Since AC voltage varies between zero and a maximum value in both positive and negative directions, it follows that the induced voltage in the secondary also varies between zero and a maximum value in both directions. Consequently, the output of a transformer is also AC unless we do something else to rectify it. In certain applications (such as automotive or charging batteries where the battery requires a DC current) the AC output of the generator is rectified using diodes. The positive polarity current goes through directly, but the negative portion is switched back into positive. Consequently, the current becomes DC.

In transformers used for increasing or decreasing voltage the primary and secondary coils are usually kept parallel, and consequently, the angle between them is zero. As a result, $\cos \beta$

is 1 and the induced voltage achieves its maximum value. However, in other applications such as *resolvers*, the angle may be changed by rotating one of the coils relative to the other, changing $\cos \beta$ and influencing the output voltage. Resolvers are used as sensors to measure the angle of rotation of shafts and joints in systems such as robots.

The output voltage is also proportional to the ratio of the primary and secondary coils as N_2/N_1. This means that if the number of turns in the secondary coil is larger than the primary coil, the output voltage is larger than the input voltage and vice versa. This is the primary application of transformers; by selecting the ratio of the number of turns in the primary and secondary coils, we can achieve any desirable voltage ratio.

As mentioned earlier, if N_2/N_1 is larger than 1, the output voltage will be increased. If it is smaller than 1, it will be decreased. Assume that we use a ratio of 10/1. This means that the output voltage will increase by a factor close to 10. Does this mean that we have increased the power of the system at no cost? Obviously if this were true, we could use a transformer to "generate" additional power indefinitely at no cost, which is impossible. So what else should we expect? Instead of considering only the voltage we must consider the power, which for electrical systems is defined as:

$$P = VI, \tag{6.4}$$

where P is the power in watts, V is the voltage and I is the current. In other words, the power of an electrical system is the product of its voltage and current. We have seen that, except for losses (which are always present, and the best we can do is to approach 100% efficiency, but never reach 100%), the power should remain the same; output power should be equal to input power because we do not generate power or energy out of nowhere. Therefore, we should expect that when we increase the voltage, the corresponding current of the system reduces proportionally, and when we decrease the voltage, its current increases proportionally. As an example, and assuming almost 100% efficiency, if we increase the voltage by a factor of 10, the current in the secondary will be 1/10th of the current in the primary. And this is exactly what a transformer does. It increases or decreases the voltage at the expense of the current. We are not changing the total power (or energy) available; we are just transforming the ratios of the voltage and current at each other's expense.

Electric power transmission is the main application of this system. To better understand this, let's first look at electrical power loss in electrical systems. All electrical conductors, even the best materials (like copper), show some resistance to the free flow of electrons. This means that as electrons move in a conductor, some of the energy they have is converted to heat energy. The power lost as heat energy in a conductor is expressed by:

$$P_{lost} = RI^2, \tag{6.5}$$

where I is the current and R is the resistance. Clearly, since I is squared, it is a much more important factor than resistance. In other words, in order to reduce power loss in conductors, it is more important and more effective to reduce the current than it is to reduce resistance. Resistance can be reduced by increasing the cross section of the conductor (thicker wires) which increases the weight and the cost of the wire, sometimes prohibitively. However, reducing the current at the same rate reduces power loss much more significantly. And this is exactly why transformers are used.

Electrical power generation is usually accomplished in power plants in locations where they make the most sense. For example, hydroelectric power plants do not require fuel such as oil or gas, generating (actually, converting the potential energy of the water behind a dam into) electrical energy at very low cost; the power comes from the kinetic and/or potential energy of the water running through the turbines. However, dams are usually not close to cities or communities where electricity is needed and consequently, the power must be transmitted. Another concern might be pollution, noise, and economics (cost of the land). In most cities it is impossible to operate a power plant within the city limits. Power plants are in the outskirts and their power must be transmitted. And perhaps most importantly, it is very uneconomical to have small power plants in each neighborhood for the consumption of the small community around the plant; large power plants are much more efficient and economical. Therefore, a few large power plants generate enormous amounts of electrical energy that are transmitted to large areas for consumption. Nowadays, almost all plants, whether hydroelectric, fuel based, or solar and wind, are interconnected through a grid which feeds all communities. Consequently, it is crucial to be able to transmit huge amounts of electrical energy from one place to another.

However, as we saw, when power is transmitted through a conductor, some of it inevitably converts to heat energy due to electrical resistance in the conductor. To reduce loss, we can either use heavy-gauge copper wires at enormous weight and cost or try to reduce the current. To understand this issue, suppose that a power plant generates electricity at 100 volts of potential and 100 amps of current, yielding $100 \times 100 = 10,000$ watts of power. In this case, the loss of power in a length of wire with 1 ohm of resistance will be $P_{lost} = RI^2 = (1)(100)^2 = 10,000$ watts, an enormous amount, basically wasting all the power that was generated.

Now suppose that using a transformer, we increase the voltage to 10,000 volts. This means that, assuming there is no loss, the current will be reduced to 1 amp, still yielding 10,000 watts. In this case, the power loss for the same length of wire equaling 1 ohm will be $P_{lost} = RI^2 = (1)(1)^2 = 1$ watt. So the loss is 1/10,000th of the first case, allowing the power to be distributed to a very large area. This means that the power at the point of consumption is at 10,000 volts and 1 amp, which is completely useless; we need high-current power at 110 volts (220 volts in many other countries) to run our machines, appliances, and devices. However, if another transformer with the opposite ratio of the first one is used to once again reverse the transformation, we can recover the same 100 volts and 100 amps at the point of consumption and deliver appropriate power to the user. This is exactly what is done. In the first part of the transmission journey over rural transmission towers and lines for long distances, the voltage may be increased to hundreds of thousands of volts with very little current (in the U.S., there are power lines with 115, 138, 161, 230, 345, and 500 kV). In sub-stations, this is reduced to tens of thousands of volts for local transmission, and finally, reduced again by local transformers in neighborhoods to 110 volts and delivered to users. Figure 6.24 shows typical transmission towers and transformer units on top of an electrical pole, reducing electric power from tens of thousands of volts to 110 volts.

Figure 6.24: Typical transmission lines and transformers on power poles that reduce electric power from tens of thousands of volts to 110 volts for domestic consumption.

However, all this is possible because we deal with AC power which provides the necessary variation in the flux density needed for Faraday's Law to work. DC power would not provide this opportunity because it does not change; to induce a voltage in the secondary coil, there would have to be a motion present which is not the case for transformers (otherwise, it becomes a generator which we will see later). Nowadays, it is possible to electronically switch on and off a DC current and cause it to induce a voltage in the secondary coil, and consequently, have a DC transformer. But this was not the case in the past.

The story is that Thomas Edison had spent a lot of money to establish local DC-generating power plants in different neighborhoods of New York City and to transmit them via copper wires to households. The first one was in 1882 on Pearl Street in lower Manhattan. However, as discussed, to reduce power loss in the wires, he had to use very thick wires for the transmission of low voltage, high current electricity. At the time when he started transformers did not exist anyway, but even if they did, they would have been ineffective with a DC current. Later Tesla designed and built prototype transformers with AC power. George Westinghouse, an entrepreneurial inventor and pioneer of Edison's era, seized the opportunity to generate low cost hydroelectric AC power at Niagara Falls, and by transforming it to high voltage and using very thin wires, transmit the power to New York City at very low cost and compete with Edison. The rivalry was intense, and at one point Edison had his engineers design and build what is now known as the electric chair (which is used for execution) in order to scare people from using AC power. However, this did not work, and Westinghouse became a huge company, at one time employing more people than any other company in the U.S.

In reality, there are three lines of transmission for three-phase power (needed for higher voltages and currents in larger plants and certain applications like three-phase motors). Each of the three phases is treated exactly the same, but carried on a separate wire. To reduce the electromagnetic effects of these high voltages, the wires are usually drawn in a braided manner (they never touch; in fact they are apart from each other far enough to prevent arching between them, but braided).

Chargers we use to recharge our batteries are miniature transformers too. Their primary and secondary coils are designed to provide the proper voltage needed. Some chargers provide DC output. This is done externally by diodes, rectifiers, and capacitors, etc. In other words, although the output of the transformer is an AC current, diodes and rectifiers eliminate or switch the negative polarity portion to a positive one, and capacitors or other averaging circuits modify the rectified output close to a DC (see Section 6.10).

In some transformers the secondary coil is tapped at different counts of turns, generating a variety of voltages which are accessible to the user. Therefore, the user may choose a variety of different voltages depending on the application. Figure 6.23 shows such a transformer.

6.9   DC GENERATORS

For most cases, a generator is nothing different than a motor. Instead of supplying a current to the motor and expecting it to rotate and provide a torque, a generator is rotated externally (by a torque) and is expected to provide a current (or voltage) as long as it is connected to a load such as a lamp. Otherwise, since it is not in a complete circuit, it will rotate without resistance and no power will be generated. This is still within the parameters of Faraday's Law; because we are changing the flux intensity by rotating the rotor, a voltage is induced in the conductors.

A DC generator is the same as a DC motor. Since we intentionally use brushes and commutators, as discussed earlier, the output of a DC generator is discontinuous and choppy. This means that although the current is always in one direction and the polarity is always the same, the current does fluctuate. However, for most purposes such as charging batteries this is not a problem. But for a true DC current that, similar to the current from a battery, does not fluctuate, the output must be smoothed.

6.10   AC GENERATORS

As with DC generators, for most cases an AC generator is also the same as an AC motor. When the rotor is rotated by an external torque, the magnetic field induces a voltage or current in the stator as long as a complete circuit exists. Therefore, when attached to a load, it can operate the device. However, in order to use an AC generator for charging batteries the current must be rectified; the reverse polarity section of the current must be switched to forward polarity (remember Figure 6.4 and how the polarity of an AC current reverses every 1/2 cycle). Simple diode arrangements called *rectifiers* are used to do this, and therefore, although the current is not constant, it can be used to charge batteries. This is the set-up used in automobiles too. The generator is usually an AC generator with rectifiers in it. Figure 6.25 shows a simple rectifier circuit made up of four diodes in the form of a bridge. In the forward polarity portion the current flows through diode B, through the load, through diode D and back to the source, forming a complete circuit. For the reverse polarity portion of the current notice how the direction of the positive/negative is now reversed. As soon as the polarity reverses, diode B no longer allows the current through. Instead, the current flows through diode C, through the load, through diode A, and back to the source. However, notice how in both cases, the direction of the output feeding through the load is the same. As a result, the output is now rectified and is always positive. Once again, notice that although the current is rectified, it is not constant. To create a constant-magnitude current that resembles the current from a battery it is necessary to remove the variation. To do this a simple circuit made up of a resistor and capacitor can be used to filter out the changes (this is called a *low-pass filter*). This simple combination of a resistor and capacitor reduces the variations to a large extent and makes the current smooth. The capacitor charges up when the voltage is higher and discharges back into the circuit when it is lower, smoothing the current. If too much current is drawn from the source this smoothness reduces and shows ripples in the output.

It should be mentioned here that since AC induction motors (squirrel cage) do not have magnets, they do not generate a voltage unless something else is done. This includes a capacitor to give an initial charge to the rotor to start the current, which, as long as the rotor turns at or above the nominal speed, will continue to generate electricity.

6.11   BACK-EMF ISSUES IN MOTORS AND TRANSFORMERS: LAMINATED IRON CORES

An interesting consequence of Faraday's Law is that it is also present in reverse even when it is undesirable. As we discussed earlier, due to the electromotive force phenomenon, when a current is supplied to the rotor of a motor, it rotates. Conversely, due to back-emf, when the rotor is turned by an external torque, the same Faraday's Law causes an induction of a current in the stator. And

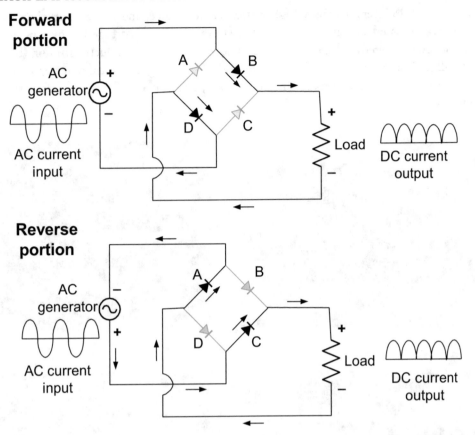

Figure 6.25: A rectifier is used to change the direction of the negative portion of the AC current into positive, thereby converting it to a DC current.

as we saw, a transformer also functions based on the same electromotive force and Faraday's Law. However, both in motors and transformers, since we need to concentrate the flux, we use a metal core or a metal rotor. Based on the back-emf and the presence of varying flux, caused by an AC current or as a result of using brushes and commutators, a current is also induced in the core of the transformer or the rotor of a motor (also called *eddy current*). If the core is solid iron, due to its low electrical resistance, the current can be large, creating a lot of heat (this is used in *induction heating*, where food is cooked when the bottom of the pot or a pan is heated by eddy currents). This is a huge waste of energy and a cause of concern to dissipate the heat. In order to reduce the effects of this back-emf in the core of the transformer or the rotor we need to reduce the flow of the back-emf current. To do so, the core of the transformer or the rotor is usually made up of thin layers of metal, laminated together, that are insulated from each other. Because the current flows in very thin layers that have higher electrical resistance, and consequently lower currents, heat generation is effectively eliminated. The layers are laminated together by pins, rivets, and welds, or pressed together. Figure 6.26 shows (a) the stator of an induction motor, (b) the rotor of a small DC motor, (c) a transformer core, and (d) the magnetic rotor of a stepper motor. All of these are made of thin layers laminated together, but insulated from each other, to form the required shape.

(a)

(b)

(c)

(d)

Figure 6.26: The rotors of motors and the cores of transformers are usually made of thin layers of the metal that are insulated from each other and connected by pins, rivets, screws, or welds to take the required shape and fight back against the induced back-emf due to the changing flux.

6.12 BACK-EMF IN DC MOTORS: SERVOMOTORS

Back-emf plays a significant role in the performance of all motors, but especially DC motors. To understand this issue first recall our discussion about the relationship between an applied torque, mass moment of inertia, and angular acceleration as described by Equation (5.16), repeated here:

$$\vec{T} = I\vec{\alpha}$$

As we discussed, when a torque is applied to a rotating body, it accelerates and rotates faster in proportion to its mass moment of inertia. As long as the torque is present and exceeds friction and other resistive torques, the body will continue to accelerate and rotate faster.

Now consider a DC motor that is connected to a battery. As the current flows through the motor and the electromotive force exerts a torque on the rotor, it will start to rotate. In proportion to the mass moment of inertia of the rotor, it will continue to accelerate and rotate faster as long as the torque is present (and of course, for lighter rotors, the acceleration is higher and vice versa). However, since this torque continues to be present, should we not expect to see the rotor's speed continue to increase, theoretically to infinity? In other words, while the current continues, so does the torque and the acceleration, increasing the angular velocity indefinitely until the rotor disintegrates. But we know from experience that if we connect a motor to a battery, when it reaches its nominal value, the velocity no longer increases. Why? This is due to the same back-emf.

Once again, let's imagine that we connect a DC motor to a battery. Since there is a torque, the rotor accelerates and its angular velocity increases. However, as mentioned before, since the rotor contains coils that are moving in the presence of a magnetic field, a back-emf voltage (or current) is induced in the coil which is in the opposite direction of the supplied current. As a result, it reduces the effective current to a value that is smaller than the supplied value. As the rotor increases its speed, the back-emf current increases and effectively reduces the supplied current until such a time that the supplied current and the back-emf current equal each other, but in opposite directions; the effective current at that speed is zero. Therefore, the effective torque at that speed goes to zero as does the resulting angular acceleration, and the velocity no longer increases. As a result, the DC motor continues to run at that nominal speed until conditions change, not increasing to infinity.

Now imagine that we engage a load, for example a fan blade or a wheel, to the motor. Since the effective torque on the rotor at the nominal speed is zero but we have added an external load, there will be a negative acceleration (deceleration) which will slow down the rotor. However, as it slows down, the back-emf will be lower, increasing the effective current, which increases the torque. Therefore, as we increase the external load on the motor, it will slow down until the torque generated by the motor equals it. At this point the motor will rotate at a constant speed that is lower than when it was not loaded. If we increase the load, the motor will further slow down to decrease the back-emf, increasing the effective current and increasing the torque supplied. This process continues as the load changes; with an increased load the motor slows down, and with a decreased load it speeds up to match the back-emf with the required torque. You may have experienced this phenomenon when dealing with DC motors, whether in appliances, toys, or other devices.

In many applications it is necessary to maintain an exact speed regardless of the load on the motor. For example, the speed of the motors of a robot arm that moves in space for welding two pieces together must be very exactly controlled; otherwise the welds will either be too thick, too thin, or non-uniform. However, the load of the arm changes as it moves through the workspace. Without a control system to maintain correct velocities it would be impossible to perform a satisfactory job. In order to control the speed or maintain a constant speed we must use a controller that, through the use of an appropriate sensor, measures the velocity of the rotor and compares it to the desired value. If the speed is lower than desired, it will increase the supplied current (or voltage) to the motor, increasing the effective current and increasing the speed. If the speed is too high, it will decrease the supplied current (or voltage) to the motor, reducing the effective current and torque and slowing down the motor. This process continues as long as the motor is running. Such a system is called a *feedback control system*. It feeds back the sensed information to the controller which compares it with the desired value, and makes appropriate adjustments as necessary. Feedback systems are not just for controlling a motor but for countless devices and systems and may take many shapes and control many other factors. But their principal intent is to control some characteristic of a device through this sensed state of affairs and to make adjustments to control the output.

A motor that is equipped with such a controller is called a *servomotor* because its velocity can be controlled. In fact, by an additional position-sensor feedback, we can easily measure how much the rotor rotates, and turn it off as it approaches the desired angle of rotation. So, the total displacement and the speed of rotation of servomotors can both be specified and controlled. But the main point of this discussion is that the back-emf continues to play a pivotal role in the way a motor responds to always-varying loads and the way it is controlled. Without the controller, the desired speeds of motors must be maintained manually.

6.13   ADVANTAGES AND DISADVANTAGES OF DIFFERENT MOTORS

Different types of motors have different characteristics, advantages, and disadvantages that make them unique in their applications and utility. In this section we will look at each type and learn about their characteristics. The major issues related to motors are heat rejection, reliability and life expectancy, torque rating, ability to reverse direction, and control of displacement and speed.

Heat rejection is an important issue in motors. As we discussed before, when a current flows through a conductor, due to the ever-presence of some resistance in the wires, heat is generated according to Equation (6.5). This heat increases as the current or the resistance increases. If not rejected or dissipated, the heat can severely damage the motor. Heat production is more prominent when the motor is under load. As we discussed, when the load on a motor increases, it slows down, the back-emf is lower, and the effective current is higher to provide additional torque. But the higher current also means higher heat production. The worst case is when the load is so high that the motor eventually stops due to the lack of enough torque, even though the back-emf is zero. This is called *stall condition*, when heat generation is at its maximum. Stall condition can burn a motor if it is not prevented.

In DC motors the current flows through the rotor coils and heat is therefore accumulated in the rotor. This heat has to flow from the rotor through the air gap into the stator, and through the stator and the body of the motor, out to the environment either by radiation (if you are close to a motor, you may feel its heat even if you do not touch it), convection (air circulating around the motor's body and taking the heat with it as it warms up), or in some cases by conduction (when the motor is attached to something else and the heat flows through it). This is a long path with high resistance provided by the air gap (air is a very good heat insulator compared to metals). On the other hand, heat generation in AC motors is in the stator because the current flows through the stator coil. Therefore, heat only has to flow through and out of the body. The heat path is much shorter and simpler as shown in Figure 6.27. Therefore, AC-type construction where the stator carries the current is more rugged, can withstand a much higher current, and consequently, produce a larger torque. As a result, AC-type motors are generally more powerful than their counterparts in DC. Notice that stepper motors have an AC-type construction where the current flows through the stator although they operate on a DC current. Therefore, they can generally handle larger currents.

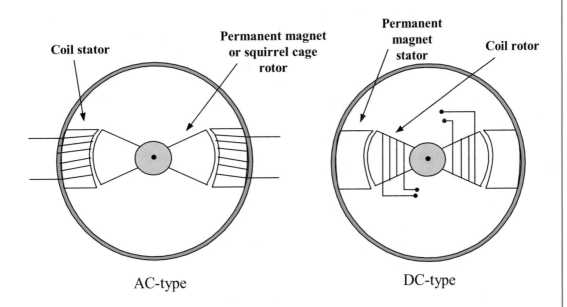

Figure 6.27: Heat rejection path for AC- and DC-type construction.

Reliability and life expectancy results from simplicity in design and using fewer parts. DC motors have more parts, including commutators, brushes, and springs. Of more concern are the brushes that wear out and need replacement once in a while. As a result, AC-type motors, steppers motors, and brushless DC motors that do not employ commutators and brushes are generally more rugged and longer lasting.

Torque rating relates to the response of the motor in relation to the supplied current. The generated torque of DC motors is nearly proportional with the supplied current. AC motors can handle larger currents and as a result can be more powerful for the same coil sizes and dimensions. However, stepper motors are generally the weakest. The largest torque they develop (called *holding torque*) is when they do not rotate at all. As they start to rotate, their torque decreases rather rapidly to the point that if they rotate fast, the torque becomes so small that the motor will miss steps. Since steppers do not usually have feedback systems, the controller will not be aware of their missed steps; this can have unacceptable consequences. The main reason for this behavior is that since the fields in the stepper coils are turned on and off very rapidly as their speed increases, the back-emf current fights the supplied current and severely affects the torque. There are remedies to minimize the effects of back-emf (such as micro-stepping and the application of zener-diodes, etc.) at an added cost.

The ability to reverse direction is a major factor. In many cases there is no need to reverse the direction of rotation of the motor (such as a fan). In that case, AC-type motors have many advantages. But when direction control is important, the choice is either a reversible AC motor or a DC motor (including universal motors). Because of this, most servomotors are of DC-type.

Displacement and speed control is another deciding factor in the choice of motors. For example, stepper motors and brushless DC motors are run one step at a time, and consequently, it is easy to count the number of steps (or actually command the controller to send a known number of signals to the motor to move the rotor an exact number of degrees) and how fast the signals are sent and therefore control the displacement and speed of the motor. Therefore, there is no need for a controller system or feedback sensors to measure the motion of the motors. However, these motors do require circuitry to operate with the added cost. Additionally, if these motors miss a step, there is no control system involved and no feedback system to determine the error and correct the motor. As a result, they can only be used for situations where there is little chance of missing a step or when they are reset often, like in a printer where the motor's position is reset at the end of every line.

DC and AC motors simply rotate as long as there is a current, and therefore, there is no control over their motion unless we employ a control system that through the use of sensors, measures both the rotational displacement and speed of the rotor and provides control over them (namely a servomotor). In this case, there is no need for drive circuitry as in a stepper motor or brushless DC motor, but there is a need for a control system. So depending on the application, the users must select the best choice based on their needs.

These characteristics are also present in other types of motors. Depending on their characteristics, you can determine their utility and application. For example, any motor that does not have a rotor through which current flows will not have to deal with heat rejection issues. Unless other means are provided, AC currents from an outlet cannot be modified, and therefore, AC-type motors cannot be reversed. So even if we have not discussed other motors here, you can determine their characteristics by comparison with these fundamental types.

Next time you see a motor or a transformer, especially if you have the chance to open it up and see its inside, see if you can determine what type of a motor it is and why it is used for that particular purpose. Electromagnetism and electromotive force are an important part of our daily lives in countless ways.

Author's Biography

SAEED BENJAMIN NIKU

Saeed Benjamin Niku is a professor of mechanical engineering at California Polytechnic State University (Cal Poly), San Luis Obispo, California. He has taught courses in mechanics, robotics, design, and creativity, and has been involved in the design of many products, including many assistive devices for the disabled and new robotic devices. Dr. Niku's publications include a statics workbook, an introduction to robotic analysis book (in its second edition), and a creative design of products and systems book. He enjoys making furniture and utilitarian products as well as artistic artifacts with wood, metals, glass, leather, and other materials.

He received a B.S. mechanical engineering from Tehran Polytechnic in 1975, an M.S. mechanical engineering from Stanford University in 1976, and a Ph.D. in mechanical engineering from the University of California, Davis in 1982.

Dr. Niku is a licensed professional engineer in the State of California.

Index

Printed in the United States
by Baker & Taylor Publisher Services